T0174847

Defining and Measuring Nature
(Second Edition)
The make of all things

Defining and Measuring Nature (Second Edition)

The make of all things

Jeffrey Huw Williams

IOP Publishing, Bristol, UK

ISBN   978-0-7503-3143-2 (ebook)
ISBN   978-0-7503-3141-8 (print)
ISBN   978-0-7503-3144-9 (myPrint)
ISBN   978-0-7503-3142-5 (mobi)

DOI   10.1088/978-0-7503-3143-2

Version: 20200701

IOP ebooks

British Library Cataloguing-in-Publication Data: A catalogue record for this book is available from the British Library.

Published by IOP Publishing, wholly owned by The Institute of Physics, London

IOP Publishing, Temple Circus, Temple Way, Bristol, BS1 6HG, UK

US Office: IOP Publishing, Inc., 190 North Independence Mall West, Suite 601, Philadelphia, PA 19106, USA

For Bruce Currie, without whose encouragement
this volume would probably not have seen the light of day.

Contents

Preface

On 9 March 1790, Charles Maurice de Talleyrand-Périgord, Bishop of Autun and *député du clergé* (a deputy representing the church) in the revolutionary National Constituent Assembly of France, presented to his fellow deputies proposals for the standardization of the French systems of weights and measures. The opportunity afforded by the changes taking place, as a consequence of the French Revolution was considered to be the appropriate moment for a reform of measurement in France.

This same Charles Maurice de Talleyrand-Périgord went on to become one of the most famous and venal of diplomats, but in 1790 he was merely an ambitious 36-year-old libertine and aristocratic bishop of the *ancien régime*. On that day, Deputy Talleyrand first suggested that the whole Kingdom of France (King Louis XVI kept his head until early-1793) be forced to use those units of measurement habitually used in Paris. After rejecting the bishop's Paris-centred solution to the lack of coherence in France's weights and measures, the deputies considered a very different proposal from him; one which was to have enormous consequences for the future of France, and for the evolution of modern science and technology.

The National Constituent Assembly went on to discuss the creation of an entirely new system of weights and measures, one based on natural phenomena and not on the size of the royal foot, or an old barrel kept in a provincial *Palace du Justice*. Talleyrand proposed the creation of the system of weights and measures known today as the Metric System. Within a year, Talleyrand had moved away from the church and had gone into politics and diplomacy, but as an unelected bishop he gave political birth to the Metric System.

This project for the creation of an entirely new system of weights and measures was approved by the National Constituent Assembly on 8 May 1790, and received the Royal Assent on 22 August. The politicians then confided the details of creating this new system of weights and measures to the *Académie des sciences* (the French national academy of sciences). The *Académie* had already begun thinking about the creation of such a natural system of weights and measures, and had recently nominated a commission of eminent French *savants* or natural philosophers (the precursors of modern scientists) to investigate the use of a variety of readily measured natural phenomena to serve as the basis for standards of length and mass.

This was the *apogée* of the Age of Enlightenment, the logic of standardization and of a quantitative understanding of the natural world were the driving forces of the *savants* of the day. France was beginning her great revolution, whose emblems and slogans still decorate public buildings in France today, and still resonate throughout the politics of our world. The French Revolution of 1789 launched the creation of the modern world, particularly our present scientific world-view. Indeed, we will see that the Metric System grew directly from the most radical period of the French Revolution; scientific revolutions often go hand-in-hand with political upheaval.

By the profound changes that the French Revolution wrought, not only on French society but on the wider world, weights and measures passed from being purely the preserve of merchants and investors, to become the essential universal language of the new profession of 'scientist' (a word coined in the social ferment of the French Revolution). These changes were the foundation upon which the triumph of Newton's mechanical model of Nature was achieved in the 19th Century. Gone were the confusing and strife-provoking local definitions of acres, pints, bushels, gallons and pounds. The new Metric System would be a language based on the dimensions of the Earth itself.

Not only would the new rational system of weights and measures be established for the citizens of the new French Republic, but the *savants* who created these units hoped that they would be exported and adopted by all humanity. Nicolas de Caritat, *marquis* de Condorcet, was an enlightened thinker in an age of great humanists. He was not a revolutionary politician, but someone who was prepared to use the opportunity of regime change to better the human condition. Condorcet was one of history's eternal optimists for human progress, and he was the permanent secretary of the *Académie des sciences*, in which position he coined the memorable slogan for the new Metric System: 'A tous les temps. A tous les peuples' (For all time. For all people).

Weights and measures form a part of everyone's ingrained mental landscapes, whether they know it or not. Indeed, it is impossible to function effectively or efficiently in the modern world without some internalized system of measurement which enables one to estimate or judge size, volume, weight, duration, length and value. Children may learn something of such systems of weights and measures in their early-years at school, and then as adults incorporate them unconsciously, and use them for the remainder of their lives.

In this volume, I shall outline a history of measurement science (metrology), and use this history to show how the language that is modern science evolved. This scientific language allows us to comprehend all the phenomena that we see in Nature (and will allow us to comprehend those phenomena that we have yet to observe) through only seven base quantities (length, mass, time, amount of electricity, temperature, light intensity and amount of substance). The history I shall present will be that of the Metric System, which is today known to the scientific community as the *Système international des unités* (International System of Units) or SI. In particular, we will look at the recent redefinitions of the base quantities, or base units that make up the SI; that is, the creation of the Quantum-SI. But before commencing, it is as well to note that as the Metric System was created in France and as the organization responsible for maintaining the modern scientific version of the Metric System is in France, the official language of all matters pertaining to the Metric System is French.

Acknowledgements

The author would like to express his thanks to Dr Janet Miles, BIPM, editor of *Metrologia* for permission to use published images, and for suggestions concerning appropriate references. Thanks are also due to the BIPM for permission to reproduce a number of photographs. In addition, the author would like to thank Dr Emilio Pisanty, ICFO—The Institute of Photonic Sciences, Barcelona, for permission to use his compact design of the interconnectedness of the base quantities in the Quantum-SI. Finally, the author would like to acknowledge the patience and assistance of the staff of the IOP Publishing in helping to bring this second edition to fruition.

Un grand merci à tous.

Author biography

Jeffrey Huw Williams

The author was born in Swansea, UK, in 1956. He attended the University College of Wales, Aberystwyth and Cambridge University; being awarded a PhD in chemical physics from the University of Cambridge in 1981. Subsequently, his career as a research scientist was in the physical sciences. First, as a research scientist in the universities of Cambridge, Oxford, Harvard and Illinois, and subsequently as an experimental physicist at the Institute Laue-Langevin, Grenoble, which remains one of the world's leading centres for research involving neutrons. During this research career, the author published more than seventy technical papers and invited review articles in the peer-reviewed literature. However, after much thought the author chose to leave research in 1992, and moved to the world of science publishing and the communication of science by becoming the European editor for the physical sciences for the AAAS's magazine *Science*.

Subsequently, the author was Assistant Executive Secretary of the International Union of Pure and Applied Chemistry; the agency responsible for the world-wide advancement of chemistry through international collaboration. And most recently, 2003–2008, he was the head of publications and communications at the *Bureau international des poids et mesures* (BIPM), Sèvres. The BIPM is charged by the Metre Convention of 1875 with ensuring world-wide uniformity of measurements, and their traceability to the International System of Units (SI). It was during these years at the BIPM that the author became interested in, and familiar with the origin of the Metric System, its subsequent evolution into the SI, and the recent transformation into the Quantum-SI.

Since retiring, the author has devoted himself to writing. In 2014, he published *Defining and Measuring Nature: the make of all things* in the IOP Concise Physics series. In 2015, he published *Order from Force: a natural history of the vacuum* in the IOP Concise Physics series. This title looks at intermolecular forces, but also explores how ordered structures, whether they are galaxies or crystalline solids, arise via the application of a force. Then in 2016, he published *Quantifying Measurement: the tyranny of number*, again in the IOP Concise Physics series. This title is intended to explain the concepts essential in an understanding of the origins of measurement uncertainty. No matter how well an experiment is done, there is always an uncertainty associated with the final result. In 2017, he published *Crystal Engineering: how molecules build solids* in the IOP Concise Physics series. This title looks at how the many millions of molecules, of hugely varying shapes and size can all be packed into a handful of crystal symmetries. Most recently, 2018, the author published *The Molecule as Meme*, again in the IOP Concise Physics Series. This title explains how the originally separate sciences of physics and chemistry became one science, with the advent of quantum mechanics and the acceptance of the existence of molecules.

In addition, retirement has allowed the author to return to the research laboratory, and he is again publishing technical papers; this time in the fields of crystal design and structure determination via x-ray diffraction; in particular, the architecture and temperature stability of co-crystals and molecular adducts.

Introduction: the origin of observation and measurement

History and archaeology tell us that when humans first began to congregate and settle, they codified their lives and experiences, and formed a collective for mutual support. This proto-civilization arose inevitably from each individual's questions about the world, and their attempt to understand themselves, each other, and their place in the world. These groups, or tribes evolved rules of conduct to facilitate communal living, and they made a calendar for the celebration of important events, such as planting the crops, and when to go hunting and fishing; events upon which the group was utterly dependent. These early tribal societies also preserved their songs, their experiences or history, and their stories, fables, wisdom and beliefs in the memories of the tribe's Shaman or Bard. These collective memories led to myths and legends, which were extravagant and hence memorable. They were short-hand records of matters such as: invasions, migrations, conquests, dynastic changes, admission and adoption of foreign religious cults, and of social reforms. This process of social evolution was also the origin of religion, and of a magical way of looking at Nature; both of which are still with us today. Eventually, this evolutionary process was also the origin of science, which is essentially our investigation of Nature to understand something of what is happening around us.

Our ancestors would have observed the progression of the Sun and the Moon; the ephemerides, so as to better understand when to undertake essential tasks. This information would first have been memorized, as literacy had yet to develop. Eventually, the long lists of observed phenomena and things would have been preserved; perhaps in painting, perhaps in carvings, or perhaps in the form of orally-transmitted poetry or myth. But how did this transformation to a scientific view-point came about? Given that Chinese civilization is our oldest and best documented record of how societies develop, evolve and became interconnected; it is in Ancient China that we must look for the most distant origins of science [1].

What was the main motive for the earliest natural philosophers, or Taoists of Ancient China, compelling them to engage in the observation and study of Nature? The answer is straightforward: to gain that peace of mind that comes from having formulated a hypothesis, however simple and provisional, about the most terrifying and poorly-understood manifestations of Nature. Nature would have been seen as being all-powerful, and indifferent to the suffering of men. Those ancient societies would have seen that the violet forces of Nature were able to easily perturb, and destroy the fragile structure of human society. Nature still has this power; as the effects of global climate change upon our engineered civilization gathers pace.

Whether the natural phenomena studied by the ancient Taoists were earthquakes, volcanic eruptions, floods, storms, or the various forms of plagues and disease, at the beginning of the adventure of science man felt himself to be stronger, more secure once he had differentiated and classified the phenomena that assailed him. This security was particularly the case when he could name those plagues and disasters. But then, 'to name is to know'; if you can name something, you have power over it—or you think you do. The origin of the name given to the newly-identified violent natural phenomenon, or plague would entail some character of the nature of that violent, destructive event. Thus, the men who first observed Nature, and named and described the observed natural phenomena would have formulated some naturalistic theory about the origin, and likely re-occurrence of those phenomena. This peace of mind, was known to the Chinese as *ching hsin*. The atomistic followers of Democritus and Epicurus in the West, knew it as *ataraxy* (calmness, or peace of mind; emotional tranquillity).

The writings known as the Zhuangzi by Zhuang Zhou; dating from the Warring States Period (476–221 BCE) tell us that 'The true men of old had no anxiety when they awoke, forgot all fear of death and composedly went and came.' These ancient natural philosophers knew, from their studies of Nature that there was an order behind the apparent violent indifference of Nature. They had observed Nature, and had seen that it is possible for man to live in harmony with Nature, and not be the passive subject of Nature's more violent manifestations. The same confidence in the power of, and the security that comes from observation and investigation is found in the West in Lucretius (c. 99–c. 55 BCE). The *De Rerum Natura* speaks of observation and deduction (that is, modern empirical science) as the only remedy for the numerous fears of mankind.

In modern science, the relationship between the rational and the empirical is seen to be so obvious as to require no explanation. However, this was not always the case. To emerge into the light in Europe, the modern scientific method had to struggle against mediaeval scholastic rationalism. Even up until the early-17th Century, the proper marriage of rational thought to empirical observations had not been consummated. At this time, it was considered in the ironic, but Taoist words of Robert Boyle, the 'father of chemistry' and author of the *Sceptical Chymist* of 1661, 'much more high and Philosophical to discover things a priore than a postiore' [2].

The origins of modern science are to be found in the interaction of four tendencies; two on one side of the argument and two on the other side. On one side, theological philosophy allied itself with Aristotelian scholastic rationalism to oppose those natural philosophers who wished to more fully understand Nature by observation. On the other side of the argument, were those natural philosophers who wished to use experimental empiricism to explore Nature. This later group would have been 'experimentalists' who reacted against formal scholasticism, and who found a powerful ally in mysticism. In the European Middle Ages, Christian theology, given its universal domination was on both sides of the argument about the rise of modern scientific methodology. But while rational theology was anti-scientific; mystical theology, or a mystical or spiritual view of the Divine, in all its aspects proved to be pro-scientific. The explanation for this apparent contradiction is to be found in the nature of magic; that essential pre-scientific element from which science evolved. Rational theology was vehemently anti-magical, but mystical theology tended to be more tolerant of magic and belief in magic. There was an affinity between mystical theology and Hermeticism in Europe, and that affinity arose from the study of Nature.

The fundamental cleavage here was not between those who were prepared to use reason to understand Nature, and those who felt reason to be insufficient to understand Nature, but between those who were prepared to use their hands and those who refused to do so; that is, between experimentalists and theoreticians or philosophers. The Vatican theologians of the Inquisition, for example, who declined Galileo's offer to look through his telescope and to see for themselves that the Ptolemaic System was incorrect, were scholastics and so they believed they were already in possession of sufficient knowledge about the visible universe. As it turned out, those Renaissance Nature mystics had discovered a new way of looking at Nature; a way which better rationalized what they saw around them.

This explanation of the origins of the modern scientific method explains so much about the 'less rational' interests of those earliest scientists. Why it was that men such as Isaac Newton, Robert Boyle and Sir Thomas Browne were interested in the Kabbalah and astrology, believing that such ancient mystical doctrines, the Hermetica contained ideas of value to them in their new empirical studies.

It may be said, therefore, that at the early stages of modern science in Europe, the mystical was often more helpful than the rationalist approach, when it came to finding an explanation or a cause of an observed phenomenon. This situation mirrors the rise of a scientific-like world-view among the philosophical scholars of Ancient China. Resting on the value placed on manual operations; that is, doing experiments rather than merely attempting to think of an explanation, and not testing that theory by experiment. Individuals such as Isaac Newton, were active laboratory workers as well as thinkers and writers. The Confucian social scholastics of Ancient China, like the rationalist Aristotelians and Thomists of mediaeval Europe nearly two millennia later, had neither sympathy for, nor interest in manual operations. Hence practical science and magic were together driven into mystical heterodoxy. It is the association of nature mysticism and empiricism that is the foundation of post-Renaissance scientific thought in the West.

There is, of course a great difference between the nature mysticism that triggered the creation of modern science (but which did not lead to the scientific triumphalism of the late-19th Century) and other forms of mysticism, which are focused in purely religious contemplation. The Ancient Chinese Taoists and the first European scientists learned that by looking at the world, and thinking about what it was they were observing, a larger number of individuals could gain a first-hand experience of the Divine; that is, of the truth.

These comments attempt to show how the first-hand study of Nature brought about the rise of modern science. This study of Nature was more than merely a philosophical exercise. We will see in this volume, how the study of the scale of the natural world gave *savants* and, later scientists the sophisticated systems of units required to better explore the more fundamental aspects of that natural world, both here on Earth and elsewhere.

Further reading

[1] In the sections of this work, where I discuss Ancient China, I will be making use of the magisterial *Science and Civilization in China* (published 1956, re-published 1975), Joseph Needham; Cambridge, Cambridge University Press. In this section, I make reference to Volume II of this multi-volume work, *The History of Scientific Thought.*

[2] *The Sceptical Chymist: or chymico-physical doubts and paradoxes, touching the experiments whereby vulgar spargyrists are wont to endeavour to evince ... of things. to which in this edition are...,* by Robert Boyle (published originally in London in 1661). There are many modern editions; for example, a version published in 2017 by the Leopold Classic Library.

IOP Publishing

Defining and Measuring Nature (Second Edition)
The make of all things
Jeffrey Huw Williams

Chapter 1

Measurement in antiquity

1.1 Man is the measure of all things

My intention is to demonstrate how man's ability to quantify and understand the world around him developed into a language, which is the basis of modern science and technology. This language of science is sufficiently powerful to allow us to make, for example, predictions about future events as in weather forecasting and modern astronomy and biology, thereby allowing us to better understand our place in Nature, enable us to construct reliable and advanced technologies, and to extend our lives. In this way, we are no longer at the mercy of the more violent aspects of Nature.

Strangely, the history of science interests more readers than the details of the science being discussed. More people can tell you something about Isaac Newton's place in the evolution of science than can list his laws of motion. But like the study of history itself, the story of the advance of science is one of the keystones of our civilization. When we know something of the origins and evolution of our assumed knowledge or understanding of Nature, we are delivered from the servitude of preconceived opinions and ideas. We better understand the limited value and shelf-life of our hypotheses. We clearly see how misunderstandings arose, and how they were resolved. And we are able to put the achievements of our own time into a more just perspective.

Any history of a subject as diffuse as measurement science (metrology) is, by necessity, a personal history of the author. The conceit that history is the essence of innumerable biographies, becomes the author's choice of which particular biographies he or she believes are needed to give his or her personal view of how our systems of weights and measures have evolved. A detailed history of the evolution and origins of the world's measurement system would fill many volumes and would probably be of limited general interest. But what I hope to demonstrate is that the incremental advances in science have driven the forward evolution of civilization. Politics may always appear to be the same, but once a scientific advance has been made, it is like a pawl and ratchet in the clockwork mechanism of history that prevents the wheels from going backwards. The advance of science inexorably drives the advance of our civilization.

Many individuals have made great efforts to improve our systems of measurement, to make them more rational, sensible, and, above all, more equitable. A positive reform of any system of weights and measures is also a reform of the abuses that ordinary people have been subjected to in their daily lives, and an end to the injuries inflicted on the weak by those with political and financial advantage. In this way, not only is technology being propelled forward, but the lives of ordinary people are made more secure, and we advance the commerce upon which our civilization rests. Curiously, it is at moments of major social change when society is experiencing rapid evolution, that there is often a call for a reform of weights and measures. Mankind prefers the status quo, and only a social upheaval provides the impetus for true change. We shall see later how the world's present measurement system arose from one of the greatest of social upheavals, the French Revolution of 1789 [1].

1.2 Seeds and cosmic forces

In all early civilizations, numbers were given supernatural significance. Numbers not only contribute to a hierarchy of units of measurement, they also give to every object measured a supplementary value, whether practical or symbolic. If we can measure something, then it must have significance, or rather it will have more significance than something that cannot be measured. Even though the act of counting has retained its simplicity, the details of the different ancient systems of measurement and units long ago faded into myth; a transition that becomes increasingly obscure when we consider that in ancient civilizations, such as Egypt and Sumeria, measurement had a magical connotation, and that he who made and enforced the measurements was either a priest or a magician, and in ancient Celtic societies he would also have been a master poet.

Measuring or quantifying something is one of our most banal, quotidian activities. Without realizing it, we use the language of measurements whenever we exchange information or trade goods, services, or money on a quantitative rather than a qualitative basis. This ubiquity of measurement makes measurements invisible; we just do not notice them anymore. Yet, even if we no longer notice our dependence on measurements, our methods of measurement define who we are and what we value.

It is interesting to note that in all ancient civilizations, the smallest unit of measurement is often the same (a first international standard). This smallest unit is usually a seed or a grain, and upon such small entities were constructed systems of measurement for units related to length and area, units of mass, and units related to the volume of dry materials (as can be seen in figure 1.1). For example, units of land area are directly related to the quantity of grain needed for sowing and the quantity harvested, and upon which the society was utterly dependent.

In Ancient Sumeria, the finger width was a unit of length equal to six seeds. In Ancient China, the measurement of length and volume was organized on a system

Figure 1.1. The seeds of the carob tree (*Ceratonia siliqua*); the scale is in centimetres. The use of the seeds of this tree for metrological purposes dates back to Ancient Mesopotamia. When dried, the seeds are hard, and the distribution of sizes and densities is narrow, which makes them an ideal metrological standard. The carob pods were also used to produce refreshing juices and desserts. The carob tree is mentioned frequently in texts dating back millennia, outlining its growth and cultivation in the Middle East and North Africa, and is mentioned in the *Epic of Gilgamesh*, one of the earliest works of literature.

based on the millet seed. In Islamic countries, the finger width is divided into five, six, or seven seeds depending on how these seeds are oriented and arranged. And in Anglo-Saxon England, the unit that anchored the system of weights and measures was based on the barley seed.

Today, science is the means by which we are able to reduce the bewildering diversity of measurements made by innumerable individuals into a manageable uniformity; all within a system of symbols, or units and mathematical equations. It does not matter whether that system of units involves describing the mass of an object in pounds, in kilograms, or *catty* the Chinese pound that is still in use throughout the Far East, or of defining the length of something in metres, in furlongs, or *chi* the Chinese foot, which is also still in use after more than three millennia, or even of defining the passage of time in seconds or lunar cycles.

What is important, nay essential for the representation of science to work, is that the choice of the particular set of units be chosen appropriately, for example, the Metric System, or the British Imperial System of Units, or the customary units of China. Then, as technology is the art of using this system of units, or symbols so as to be able to predict, control, and organize future events; your flight will take off next week and will convey you safely to your destination. The scientist views everything through the refractory medium of a system of symbols or units, and technology is the handling of materials and the use of phenomena in ways that have been predicted by that system of symbols or units. Thus, an education in science and technology is essentially the training of new adepts in the use of that system of symbols and units. Hence, the importance of metrology.

1.3 The Bronze-Age

For the first agricultural societies, reliable and reproducible measurements were essential for their survival, their evolution, and stability. For trade to flourish and for the towns and cities to prosper, a system (or systems) of weights and measures was (were) needed. Otherwise, there was always the possibility that some individuals would be prepared to use the confusion arising from multiple systems of measurement to gain commercial advantages by cheating. The problem of maintaining a reputation for the just application of weights and measures is still with us today, irrespective of what our units are now called.

The ancient world may have been a very different place from our world; however, our distant ancestors also based much of their metrology on observations of Nature. In ancient Mesopotamian society, the year was divided up according to the phases of the Moon. And it was, therefore, the Moon deity (*Inanna* in Sumeria, or *Sin* in Akkadia) and whose temple was located in one of the oldest of cities, *Ur*, who was responsible for time metrology and for attributing the measurement of time to men.

The moon was thus the first astronomical measurement standard. It is a body which over a short period grows, shrinks, and disappears; it has a rhythm that is readily followed and predicted. Indeed, the Indo-European proto-word for moon, *me*, gives rise to a number of modern words related to, and including, metrology. In the same manner in which the Moon presides over measurements, she is without doubt the source of our modern units of time. The Sun is another astronomical measuring device. It is the universal measure of the length of the day, and of the various sub-divisions of the day. The Moon and the Sun not only control and regulate our environment, but they also control the temporal rhythms experienced by man.

1.4 Ancient time metrology: the calendar

The origin of our modern calendar, the Babylonian calendar is a lunisolar calendar with solar years consisting of 12 lunar months, each beginning when the New moon was first sighted low on the western horizon at sunset[1], plus an intercalary month inserted as needed by decree (sometimes there are 13 lunar months in a solar year).

The Babylonian year began with the appearance of the first New moon after the Spring equinox; that is, the day after the Sumerian goddess *Inanna*'s (also known as *Ishtar*) return from the underworld. In the Babylonian calendar, the chief deity of the Assyrians was assigned the intercalary month, demonstrating that the calendar originated in Babylon, but about the time when the Assyrian empire was absorbed into the Babylonian empire (c. 650 BCE). During the 6th Century BCE Babylonian captivity of the Jewish people, the names of the Babylonian months were adopted into the Hebrew calendar. As is often the case, when one civilization is conquered by

---

[1] This is exactly the manner in which the modern Muslim Calendar is defined. The month begins when the New moon is first sighted at the Holy City of Mecca. Islam, like Judaism, is based on a lunar calendar; Christianity is based on a solar calendar, as it evolved from the Pagan religions of the Classical World. A lunisolar calendar is one, which indicates both the phase of the Moon and the time of the solar year.

another, the gods and myths of the older civilization are often absorbed by the more powerful civilization. The Romans were the masters of this form of syncretism.

The great problem faced by the first time-metrologists, that is, the priests and scholars of the ancient world, was the different behaviour of the two natural chronometers. The Moon completes a cycle in a synodic month (about 29.53 days) and was humanity's first precision chronometer. The Sun, however, completes a cycle on a longer time scale, making it a less-precise chronometer. A calendar year is an approximation of the number of days of the Earth's orbital period as counted in a given calendar. The Gregorian, or modern calendar presents its calendar year to be either a common year of 365 days, or a leap year of 366 days (Julian calendar).

A sidereal year is the time taken by the Earth to orbit the Sun once, with respect to fixed stars. Hence, it is also the time taken for the Sun to return to the same position, with respect to the fixed stars after apparently travelling once around the ecliptic; it equals 365.256 363 004 Ephemeris days. The sidereal year differs from the tropical year; that is, the time required for the ecliptic longitude of the Sun to increase by 360°, due to the precession of the equinoxes. The sidereal year is 20 min 24.5 s longer than the mean tropical year at J2000.0 (365.242 190 402 ephemeris days) [2].

Before the discovery of the precession of the equinoxes by the ancient Greek astronomer Hipparchus (c. 190–c. 120 BCE), the difference between the sidereal and tropical years was unknown. For naked-eye observation, the shift of the constellations relative to the equinoxes only becomes apparent over many centuries, and pre-modern calendars such as Hesiod's *Works and Days* (dating from c. 700 BCE) would give the times of the year for sowing, harvest, and so on by reference to the first appearance of stars, effectively using the sidereal year. The South and Southeast Asian solar New Year, based on Indic influences is traditionally reckoned by the Sun's entry into Aries and thus the sidereal year, but is also supposed to align with the Spring equinox, and has relevance to the harvesting and planting season and thus the tropical year.

The problem for the first metrologists was how to relate the incommensurate periods of the Lunar cycle and the Solar year. Normally, there are 12 lunations in a Solar year, but sometimes there are 13. To relate the two different cycles, the calendar was calculated over a number of years. Until the 5th Century BCE, the calendar was fully observational, but beginning about 500 BCE the months began to be regulated by a lunisolar cycle of 19 years. Although usually called the Metonic cycle, after Meton of Athens (lived in Athens in the 5th Century BCE), it is likely that Meton probably learned of the cycle from the Babylonians.

The traditional Chinese calendar (another lunisolar calendar) was developed between 771 and 476 BCE, that is, during the Spring and Autumn period of the Eastern Zhou Dynasty. Before the Zhou Dynasty, only solar calendars were used. The Zhou calendar set the beginning of the year on the day of the New moon before the Winter solstice. After the first emperor, Qin Shi Huang had unified China under the Qin Dynasty in 221 BCE, and the Qin calendar was introduced. It followed most of the rules governing the earliest of calendars. The Qin calendar was used into the Han dynasty. But, Emperor Wu of Han (141–87 BCE) introduced a number of

Table 1.1. Some ancient Mesopotamian units of measurements.

Ancient unit	Modern name	Relationship	Modern metric equivalent
she	Grain seed		2.7 mm
ubâna, shusi	Finger	6 *she*	1.65 cm
ammatu, kush	Cubit	30 *shusi*	49.5 cm
kânu	Cane	6 *kush*	2.97 m
ninda	Rush	2 *kânu*	5.94 m
ush	Half-cord	60 *kush*	29.7 m
ashlu	Cord	10 *ninda*	59.4 m
beru, dana	League	180 *ashlu*	10.7 km

reforms; his Taichu Calendar ('grand beginning calendar') defined a solar year as 365 (385/1539) days, and the lunar month was 29 (43/81) days [3].

Apart from time metrology, the units of which we still use today, Ancient Mesopotamia gave us our earliest system of weights and measures. The basic unit of length was the cubit (in French *coudée*, literally 'elbow room'), which was the length of a man's forearm. The length of the standard Mesopotamian cubit can be found by consulting surviving inscribed rulers, which tell us that the actual value of the cubit varied from one town to another (table 1.1 gives the names and equivalents of some of these Ancient Mesopotamian units).

The ancient Mesopotamians were famous for their ability as astronomers and astrologers; the two words were then synonymous (it was the philosophers of Ancient Greece who first started distinguishing between these two ways of studying the night sky). The Mesopotamian civilizations lasted for millennia. Various empires rose and absorbed the small Mesopotamian states, then those empires withdrew and fell, yet the civilization and the culture of the people between the two great rivers continued and became increasingly sophisticated. Mesopotamia possessed the greatest known system of irrigation, which is what had permitted their civilization to prosper and survive. On a clay tablet describing repair work on one particular irrigation canal, we read 'The drainage trench was dug to the length of four leagues and two hundred and sixty rushes; at Ur they build for eternity' (*Inscriptions royales sumériennes et akkadiennes*, Paris 1971, p 137). This was not a bad judgement when you remember that the ancient Mesopotamian irrigation system survived until 1258 AD when it was destroyed by the invading armies of the Mongols under a grandson of Genghis Khan[2].

To establish reliability of weights and measures, and thus the maintenance of a stable, orderly society the early books of the Old Testament constantly refer to the importance of sound metrology. These texts were written in the late Bronze-Age,

---

[2] We should never underestimate our Bronze-Age ancestors' ability to build on a large scale; they had the technology to construct vast irrigation systems and the Pyramids.

perhaps as far back as 1000 BCE. Here are two examples of the importance of sound metrology taken from the Authorized Version of the Bible.

Exodus **16**:36

'Now an *omer* is the tenth part of an *ephah*.'

In the Book of Exodus, Moses receives the Torah from God and interprets the Lord's Commandments to the People of Israel. We have here an early demonstration of a preference (Divine preference?) for a decimal system of units.

Leviticus **19**:35–36

'Ye shall do no unrighteousness in judgement, in meteyard, in weight or in measure. Just balances, just weights, a just *ephah*, and a just *hin*, shall ye have: I am the Lord your God, which brought you out of the land of Egypt.'

In other words, do not use dishonest standards when measuring length, weight, or quantity or Divine wrath will overtake you. (An *ephah* was a dry measure, and a *hin* was a liquid measure.) The Book of Leviticus gives instructions to the People of Israel on how to pray and how to live, providing detailed moral and ritual laws.

Similar proscriptions may also be seen in the Koran, which was written between 610 and 632 AD; for example: 'Woe to those who give short weight! Who when they measure against others take full measure; but when they measure to them or weigh to them, diminish!' (Koran, Sura 83).

The lack of a single, universal yardstick for measurement was not a serious drawback for the earliest city-based civilizations. Measurements were made by one craftsman completing one job at a time, rather than manufacturing a number of articles piecemeal to be assembled later. It did not make much difference how accurate those ancient measuring sticks were. What was important was that everyone in that society understood the same thing when referring to each unit of measurement.

The cubit of the ancient world was the distance from the tip of the elbow to the end of the middle finger, presumably of a carpenter's or builder's arm. This distance was useful, because it was readily available and convenient. However, it was not a universal fixed dimension or standard, as builders came in all shapes and sizes between Egypt and China. But it was a near-universal measurement standard for length in the ancient world.

Once the ancient builders had started using their arms and feet for measuring distance, it was only natural that they also thought of using fingers, hands, and legs. These early metrologists quickly discovered that there are some surprisingly stable ratios in, apparently disparate, body measurements. What is now called an inch was originally the length of part of a man's thumb (*pouce* is the French word for thumb and for inch). Twelve times that distance are roughly equivalent to the length of your foot. Three times the length of the foot was roughly the distance from the tip of a man's nose to the end of the longest of his outstretched fingers. This distance closely approximates to the yard. Two yards were a fathom which was the distance between the tips of the longest fingers of a man's outstretched arms. Half a yard was the 18-inch cubit, and half a cubit was called a span, which was the distance across the hand from the tip of the thumb to the tip of the little finger when the fingers were spread out as far as possible.

Given that the human body has such in-built metrological possibilities, it was perhaps not surprising that human morphology was widely studied. Indeed, the pre-Socratic philosopher Protagoras (c. 490–c. 420 BCE) said: 'Man is the measure of all things, of things that are, that they are, of things that are not, that they are not.' This idea was revolutionary at the time and differed from other philosophical doctrines which claimed that the Universe was based on something outside human influence. What this attention to the dimensions and variability of human morphology also created was a concept, which today we know as the Golden Ratio or Golden Mean, which is sometimes attributed to Pythagoras (c. 570–c. 495 BCE). The harmonious proportions derived from a consideration of the Golden Mean in the perfect human physique are to be seen in the image known as the Vitruvian Man by Leonardo da Vinci, an illustration of the naked male body inscribed within a circle and a square.

The Ancient Chinese used (and the modern Chinese still use) a foot as their fundamental measure of length. And given the great antiquity of Chinese Civilization, and the detailed records which have survived, we know that the dimension of the standard Chinese foot showed a continuing tendency to increase in length over the 3000 years which separate the Zhou and the Qing Dynasties. The sinologist, Joseph Needham, has pointed out (*Science and Civilization in China*, 1958, volume III, p 82) that the length of the ancient Chinese unit of measurement, the foot, increased from 0.195 to 0.308 m over the last three millennia.

This observation may just be an example of unit inflation due to social pressure; however, given the length of the period involved, it may have a more significant origin. Over the last 3000 years, one of the most important factors driving human physical and social development has been the evolution of agriculture. Today, humans are, on the whole, living longer and more healthily than previously. The quality, quantity, and availability of food have increased for all mankind. Thus, the observed increase in the length of the standard Chinese foot over the last three millennia may reflect the continually increasing size of the average Chinese man or woman (the ancient Chinese used a unit of a standard male foot and a unit of a standard female foot); they were living longer and growing taller. The observation of Needham is not simple unit inflation; it is an example of the influence of social evolution on measurement science. (Table 1.2 contains some length units, with modern equivalents from Ancient Egypt. A ceremonial cubit rod may be seen in figure 1.2.)

Egyptian gods had made different arrangements with ancient Egyptians about time metrology than *Inanna* and *Sin* had made with ancient Mesopotamians. The

Table 1.2. Some ancient Egyptian units of measurement.

Ancient unit	Modern name	Relationship	Modern metric equivalent
Djebâ	finger		1.8 cm
Shesep	hand	4 *djebâ*	7.5 cm
Meh	cubit	7 *shesep*	52.3 cm
nouh, keh	cord	100 *meh*	52.3 m
Iterou	river	200 *nouh*	10.46 km

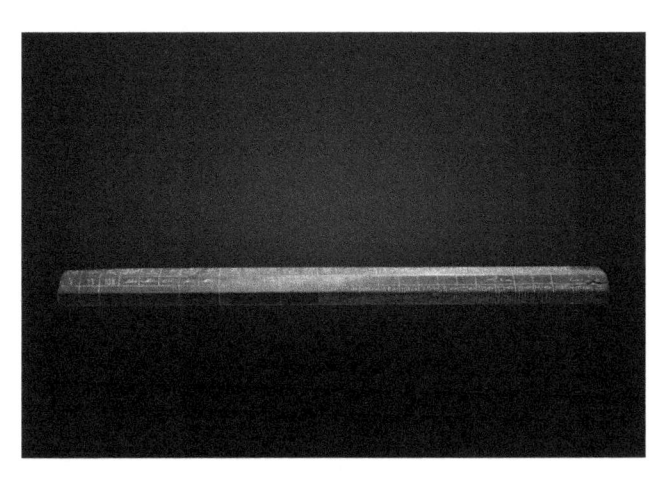

Figure 1.2. The ceremonial cubit rod of Maya (1336–27 BCE, eighteenth dynasty), an important figure during the reign of Pharaohs Tutankhamun, Ay and Horemheb of the eighteenth dynasty of Ancient Egypt. (Image from: https://en.wikipedia.org/wiki/Cubit#/media/File:Measuring_ruler-N_1538-IMG_4492-gradient.jpg; it has been obtained by the author from the Wikimedia website where it was made available by Rama under a CC BY-SA 3.0 FR licence. It is included on that basis. It is attributed to Rama.)

Egyptian year was considered to be 365 days in length. There were still 12 months each of 30 days, but the remaining 5 days were given over to religious festivals.

Some of the variation in units within a particular civilization also arose from various modes of taxation. We know that in Ancient Egypt, there was a royal cubit (about 523.5 mm) and an ordinary cubit of about 500 mm. The Pharaoh bought goods using the royal cubit and, subsequently, sold them using the common cubit; the difference between these two measures was equivalent to the amount of tax paid to the Pharaoh, that is, to the central authority of the state. The royal cubic was subdivided into seven palms of four fingers each, giving 28 fingers. This measurement standard was used from about 2700 BCE. The Great Pyramid of Giza was build using 440 royal cubits (about 230 m) along each of the four sides of its base. Many examples of actual cubit rules and measuring sticks have survived, and some of these are elaborate ceremonial rulers that were preserved in temples (see figure 1.2). Indeed, we also know that these yardsticks had to be regularly calibrated against each other on pre-set dates in the Lunar Cycle, and that the penalties for not undertaking such regular calibrations were often fatal for the metrologist-priest.

1.5 The Roman Empire

Given the influence of the Roman Empire on our modern civilization, it is not surprising that it was with the Romans that the first significant advances in the evolution of international systems of weights and measures took place. In 46 BCE, Julius Caesar (100 BCE–44 BCE), Dictator for Life of the Roman World, introduced a new Solar Calendar with three years of 365 days followed by a leap year of 366 days (the Julian Calendar).

At the same time that he set about reforming the calendar, Caesar also approved the method for measuring distances using a unit of 1000 paces, where each pace consisted of two steps. In Latin, 1000 paces is written as *mille passus*, and much later, this was shortened in English to mille and eventually to mile (*mile* in French). The Roman mile consisted of 1000 paces, where the pace or double-step was defined as five Roman feet, so the Roman mile was 5000 Roman feet. Many European countries retained a mile of 5000 feet, but in Tudor England the mile was redefined as 5280 feet in an attempt to coordinate the foot and mile with other local, non-Roman (Anglo-Saxon) units such as the rod and the furlong.

Assuming that there were five Roman feet to a Roman pace, and 1000 Roman paces to a *mille passus*, then a Roman mile would be about 1480 m. Thus, estimating the Roman mile at about 1500 m, each Roman Legionary would have had to make single paces of about 750 mm. Caesar simply selected a pace length that was a little stretched yet still comfortable for the smallest of his soldiers ... and the taller soldiers just keep in pace. Interestingly, modern armies still use a standard pace of 750 mm. Every country that was influenced by the Roman Empire, either directly by conquest and occupation or by association had units with the names inch, foot, and mile, but with local variations in length.

For measuring mass, the Romans used scales called *Libra Pondo*. In Latin, *libra* meant scales and *pondo* referred to the weights that were placed on one of the pans of the scales. The English word pound is derived from the second part of *Libra Pondo*, and the abbreviation for pound (lb) is derived from the first part.

Pounds, ounces, and hundredweights were other Roman measures of mass. The word ounce is derived from the word for a 12th part, that is, *unciae*. It is interesting that some people still use a form of *unciae* as the word inch, and it still means one-twelfth. In Rome, there were 12 ounces (*unciae*) in a pound. This survived into modern times as the 12 ounces in a Troy pound, and the 12 inches in a Roman foot; *unciae* became both inches and ounces in English.

In some parts of the English-speaking world, there are still two different pounds in everyday use: the *avoirdupois* pound and the Troy pound. The *avoirdupois* pound is divided into 16 *avoirdupois* ounces and the Troy pound is divided into 12 Troy ounces. An *avoirdupois* pound (of about 453 g) is exactly 175/144 Troy pounds. The word *avoirdupois* comes from the French phrase *avoir du poids* (to have weight), as it refers to goods sold by weight rather than by volume or piecemeal. The Troy pound is named after the French market town of Troyes, and one Troy pound (of about 373 g) is 144/175 of an *avoirdupois* pound.

Further reading

[1] For the evolution of systems of weights and measures from the Ancient World to the Metric System, the reader is referred to *Le nombre et la mesure* (Numbers and measurement) by Franck Jedrzejewski, published by Diderot Multimedia—EDL, 1999. This is a competent single volume history of the origins of the Metric System in French.

[2] *Astronomical Almanac for the Year 2017*. Washington, DC and London: United States Naval Observatory, HM Nautical Almanac Office, 2016, p M19. This publication contains a wide-

variety of both technical and general astronomical information. The book is a worldwide resource for fundamental astronomical data and the flagship publication of the Nautical Almanac Office at the US Naval Observatory and Her Majesty's Nautical Almanac Office at the UKHO including contributions from astronomical experts worldwide. It contains positions of the Sun, moon, and planets to milli-arcsecond precision, positions of minor planets and planetary satellites for each year together with data relating to Earth orientation, time-scales and coordinate systems. Phenomena including eclipses of the Sun and moon, sunrise/set, moonrise/set, and twilight times are provided as well as fundamental astronomical reference data for stars and stellar systems, observatories and related astronomical constants and techniques

[3] For all aspects of the evolution of science in China, the reference is: *Science and Civilization in China*, Joseph Needham, published by Cambridge University Press, Cambridge; particularly, volume II (*The history of Scientific Thought*) published in 1956 and volume III (Mathematics and the Sciences of the Heavens and the Earth, section 20) published in 1959.

Chapter 2

Measurement in the early modern period

2.1 'Measured by the King's iron rod'

The period following the collapse of the Roman Empire in the West is generally thought of as being dark and uncivilized. But this epoch is obscure only because of a lack of documentation; most of the science and technology known to the Romans was preserved, but it did not evolve in a systematic manner. Even after the Fall of Rome, commerce continued to flourish and new trade routes opened. Waves of supposed barbarians invaded the old Roman Empire, and there was a mixing of Roman weights and measures with those systems of measurement which existed outside the Empire. These non-Roman measures should not be thought of as being in anyway inferior to the weights and measures used in the Latin-speaking world—they were just different, but had evolved in much the same manner and for much the same, social reasons as had the units of measurement in the Mediterranean world.

The Middle Ages are marked by a tradition of Royal Decrees attempting to reform the different systems of weights and measures used in the various kingdoms. King Richard I of England, for example, proclaimed an Assize of Measures and had measurement standards in the form of iron rods distributed to major market towns throughout his domain (the Angevin empire stretched from Hadrian's Wall to the Pyrenees). The expression 'measured by the King's iron rod' appears frequently in subsequent records as an appeal to some independent, time-hallowed standard of length.

King Richard's younger brother, John was famously forced by his barons to sign a document in 1215 containing a list of fundamental rights that each of the king's subjects would possess in perpetuity, and which could not be infringed upon by arbitrary executive power. This document, *Magna Carta*, which among other things had a clause (number 35) that provided for a uniform system of weights and measures: 'There shall be one measure of wine throughout our whole realm, and one measure of ale and one measure of corn—namely, the London quarter; and one width of dyed and russet and hauberk cloths—namely, two ells below the selvage.

And with weights, moreover, it shall be as with measures.' In principle, the standardization of weights and measures was now enshrined in law, and they were not subject to the varying physical statures or whims of successive monarchs; however, the balance of authority between the king and commons was still to be resolved.

At this time, even within small regions of Europe, the various traditional means of making measurements, particularly those measurements made for purposes of comparison and law-enforcement, varied widely according to the place of measurement and the thing being measured (wheat, wine, cloth, or the surface area of arable land). Such variability in the stability of units of measurement undoubtedly arose because of the instability and flux of society. In such a society, the payment of taxes or licence fees by the peasant agricultural workers to their overlords would have encouraged the use of units of measurement which were smaller than those already in use; thereby minimizing future payment. Conversely, the overlords wanted to, at least, maintain a standard of measurement and generalize the use of an already accepted standard. Of course, the lords and *seigneurs* may have also been tempted, from time to time, to inflate the usual measurement standard and thus extract more from the workers. There were thus two powerful forces working to continually change the size of the various measurement standards in use.

Systems of weights and measurements were also, and still are stakes or bargaining chips in political games of power. Kings and feudal lords quickly learnt that it was to their advantage to possess the biggest standard unit of a particular measure and to force their measurement systems onto others. And it did not matter how this enforcement was accomplished. The measurement itself was not important; if you could control the metrology of your conquered enemies, your neighbours and their vassals, you could control their commerce and maintain your own ascendency. Measurement science was thus incorporated into the Feudal System, and the political power conferred by control of weights and measures centralized in the hands of the king, for him to distribute as he saw fit to his vassals. This, mediaeval political concept is still with us. It is the reason why the major world powers of the late-19th Century, the UK and the USA never openly accepted the Metric System from France. The UK and the USA made compromises towards the use of the Metric System, but never openly accepted to become party to a system of weights and measures based on standards kept in France (see section 3.4).

2.2 Measuring the world

One of the biggest changes to the lives of Europeans and European colonists in the 16th Century occurred in February of 1582, when Pope Gregory XIII reformed the solar calendar[1]. This long-needed change should have been instantly accepted throughout the Christian world, but as the Reformation had already splintered Christendom, the various nations adopted the new calendar in a piecemeal manner

---

[1] The reason why the Church was so concerned about the reform of the calendar was the drift in the Spring equinox. In 46 BCE, Caesar had set 25 March as the date of the equinox, and this date was no longer realistic after 16 centuries. This drift was a problem, because the central Christian feast of Easter is calculated each year in such a way that Easter Sunday is the first Sunday after the first Full moon on, or after the Spring equinox.

based on politics; England and her colonies did not adopt the change until 1752 and Russia only accepted the changes in 1917. To remedy the distortion of the solar calendar arising from the imprecision of the Julian Calendar, an Italian *savant* Aloysius Lilius (1510–76) devised a new calendar with new rules: every year that is exactly divisible by four is a leap year, except for years that are exactly divisible by 100, but the centurial years that are exactly divisible by 400 are still leap years.

The 16th Century was also notable for the introduction of a new idea to simplify everyday arithmetical operations; the use of decimal numbers (numbers to the base ten). In 1584, the Flemish engineer Simon Stevin (1548–1620) published a set of tables for the calculation of the amount of interest that banks would charge for lending money at various rates, and over various periods of time. As he was preparing these tables, Stevin realized that decimal numbers would greatly simplify calculations in every area of life.

However, it was in Italy that the greatest scientific advances were being made at this time by the natural philosopher Galileo Galilei (1564–1642). While attending Mass in the Cathedral of Pisa, he allowed his attention to wander, and it was while contemplating the swaying motion of the heavy chandeliers suspended by fine chains from the high ceiling of the cathedral that he formulated several ideas about the pendulum. Consequently, Galileo conducted detailed experiments on pendulums, thereby inventing the principle of scientific investigation, and determined the length of a pendulum swinging through its arc in exactly 1 s in Pisa. This became known as a 'seconds' pendulum'[2]. It was soon appreciated that a pendulum could be a precise measurement standard for both time (pendulum clocks) and length.

2.3 The pendulum: the world's first precision measuring device

At its simplest, a pendulum is merely a heavy weight or 'bob' suspended, at the end of a long thin wire from a fixed pivot, so that it may swing freely. When the bob is displaced from its static, vertical, or resting equilibrium position so that it is higher or further from the centre of the Earth, gravity will exert a pulling force to return the bob to the initial resting equilibrium or vertical position. But, by the time that the bob reaches its starting position, the bob is moving and momentum causes the bob to move through the initial, vertical position and it swings upwards until gravity exerts a restoring force to bring it back. These competing forces of momentum and gravity cause the bob to oscillate about the original equilibrium position; that is, the pendulum is observed to swing back and forth.

The time for one complete cycle of the bob, a left swing and a right swing, is called the period of the pendulum. If you observe the swinging motion or oscillations of the pendulum, you will see that the arc described by the motion of the bob is constant over the time-scale of most observations. This apparent, constant period of oscillation was used soon after its discovery as a means of measuring the force of gravity that Isaac Newton had announced was all pervading, and held the Universe

---

[2] It was subsequently shown that a seconds' pendulum varies in length according to where it is on the Earth's surface. At the Equator, a seconds' pendulum is 991.00 mm long, and at 45° north of the Equator, it is 993.57 mm, this difference arising because of variation in the local value of gravity.

together. However, as with most things in physics, there are qualifications to describing a pendulum as a simple harmonic oscillation. For a pendulum to behave as a perfect gravity pendulum, the pivot from which the weight is suspended must be frictionless, and the wire suspending the bob must not have any mass, and not be able to stretch or deform. In addition, the bob should not suffer any drag or fictional force as it moves through the air. These conditions of perfection are impossible to achieve under everyday conditions, and eventually all pendulums will run down naturally, and the bob will eventually return to its static position.

The pendulum is a simple measuring device. The period, T, in seconds of a pendulum of length, L (in metres) is given by $T = (2\pi/\sqrt{g})\sqrt{L}$, where g is the local value of the acceleration due to gravity [1]. The value of g varies by a few percent over the surface of the Earth, and the pendulum is a sufficiently precise device that it is capable of determining the spatial variation of g. The relatively simple relationship above yields the approximation, $T \approx 2\sqrt{L}$. Thus, the period of oscillation of a pendulum is independent of the mass of the bob. This surprising finding from the 17th Century linked the pendulum, in the imaginations of Occultists and those interested in the Hermetic arts with some, as yet undiscovered level of existence; which could well be the much sought-after link between the physical world and the world of the spirit.

The law governing simple harmonic motion was first stated by the British *savant* Robert Hooke (1635–1703) in 1660 as *Ut tensio, sic vis* or *As the extension, so the force*. However, if the amplitude of the oscillations is large, then this harmonic approximation to the motion of the pendulum cannot be made, and the equations governing the motion of the pendulum are more difficult to solve, and there is no simple relationship between the length of the pendulum and gravity. For swings of small amplitude, the period or frequency of swing is approximately the same for different size swings; that is, the period is independent of amplitude. This property, termed isochronism, is the reason pendulums are so useful for timekeeping and was first identified by Galileo. Successive swings of the pendulum take the same amount of time. Galileo first employed free-swinging pendulums in simple timing applications, such as an early metronome for musicians, and a friend used it as a time-piece to take a patient's pulse. In 1641, Galileo produced a design for a pendulum clock. However, the first true pendulum clock was built in 1656 by the Dutch *savant* and inventor Christiaan Huygens (1629–95). The precision of this first pendulum clock was a great improvement over existing mechanical clocks; increasing precision from about ±15 min per day to around ±15 s per day in the new pendulum clocks. With the invention of temperature-compensated pendulums in 1721, errors in precision pendulum clocks fell to a few seconds per week. The world was then ready for the advances in international time-keeping and surveying, which would characterise the 18th Century [2].

During his expedition to Cayenne, French Guiana (northeast coast of South America), in 1671, the French explorer and astronomer Jean Richer (1630–96) found that a pendulum clock was slower by two and a half minutes per day at Cayenne than the same pendulum clock at Paris. From this he deduced that the force of gravity was lower in French Guiana. In 1687, Isaac Newton in his *Principia Mathematica* explained that this variation in the force of gravity arose because the Earth was not a true sphere, but slightly oblate (flattened at the poles and bulging at the equator). This imperfection in the Earth's roundness coupled with the effect of centrifugal force due to its rotation, caused gravity to increase with latitude. As a consequence, portable pendulums began to be taken on voyages to distant lands, as precision gravimeters to measure the value of gravity at different points on the Earth's surface. These measurements were subsequently used to construct an accurate model for the shape of the Earth.

For three centuries, until the development of the quartz oscillator clock in the 1930s, the humble pendulum was the world's standard for accurate timekeeping. In addition, the pendulum was the first precision measuring device for a number of phenomena:

- Atmospheric pressure.

  The presence of air around the pendulum affects the period of the bob. By Archimedes' principle, the effective weight of the bob is reduced by the buoyancy of the air it displaces, while the mass (inertia) remains the same, reducing the acceleration and increasing the period.
- Gravity

  Pendulums are affected by changes in the force of gravity, which varies at different locations on Earth. Even moving a pendulum clock to the top of a tall building can cause it to lose measurable time due to the reduction in gravity.
- The seconds' pendulum.

  By the end of the 17th Century, the length of the seconds' pendulum became the standard tool for measuring gravity. But such a pendulum could also be used as a standard for defining length as the local value of gravity is proportional to the length of the pendulum, and by 1700 the length of such a pendulum had been measured with submillimetre accuracy at several cities in Europe.

Even though a pendulum is capable of giving precision information on a number of physical phenomena, a single pendulum cannot provide information on all these phenomena. In experimental science, if you have one unknown quantity to identify, you require, at least, one experimental measurement to determine the unknown value. Then, if you have two or three unknown quantities, to identify them all you require, at least, three independent experimental observations.

The Foucault pendulum was conceived by the French physicist Léon Foucault in the middle of the 19th Century, with the idea of demonstrating the rotation of the Earth, through the effect of the Coriolis force. In essence, the Foucault pendulum is a pendulum with a damping rate, sufficiently long that the precession of its plane of oscillations can be observed after, typically an hour or more. It is therefore different from the seconds' pendulum discussed above.

The Coriolis force, responsible for the precession of the Foucault pendulum, is not a force per se. Instead, it is a fictitious force which arises when problems are discussed in non-inertial frames of reference; that is, in coordinate systems which accelerate, such that the law of inertia (the derivative of momentum (p) with respect to time (t), $\mathrm{d}p/\mathrm{d}t$ = a constant) is no longer valid. In rotating systems, the two fictitious forces that arise are the centrifugal and Coriolis forces. The centrifugal force cannot be used locally to demonstrate the rotation of the Earth because the 'vertical' in every location is defined as the combination of gravity and centrifugal forces. If we wish to demonstrate dynamically that the Earth is rotating, one should consider the Coriolis effect.

At the Poles, the plane of oscillation of a Foucault pendulum remains fixed relative to the distant masses of the Universe while the Earth rotates underneath it; taking one sidereal day to complete a rotation. So, relative to Earth, the plane of oscillation of a pendulum at the North Pole—viewed from above—undergoes a full clockwise rotation during one day; a pendulum at the South Pole rotates counter-clockwise. When a Foucault pendulum is suspended at the equator, the plane of oscillation remains fixed relative to Earth. At other latitudes, the plane of oscillation precesses relative to Earth, but more slowly than at the pole; the angular speed, ω (measured in clockwise degrees per sidereal day), is proportional to the sine of the latitude, φ: $\omega = 360 \sin \varphi$/sidereal day, where latitudes north and south of the equator are defined as positive and negative, respectively.

Leon Foucault made his most famous pendulum when he suspended a 28 kg brass-coated Lead bob with a 67-m long wire from the dome of the Panthéon, Paris. The period of the pendulum was 16.5 s. Because the latitude of its location was $\varphi = 48°52'$N, the plane of the pendulum's swing made a full circle in approximately $23\mathrm{h}56'/\sin \varphi = 31.8$ h (31 h 50 min), rotating clockwise approximately 11.3° per hour. (There is a video of the Foucault pendulum in the Panthéon, Paris, at https://www.youtube.com/watch?v=59phxpjaefA.)

Although Galileo's daydreaming in church may well have contributed to the problems he had later in life with the Catholic Inquisition, his musings gave us a precision scientific tool, which is still in use today. Indeed, one could go further and say that the fame and widespread use of the pendulum in the 17th Century accords perfectly with that century's obsession with the mixing of magic and science. The swinging pendulum is an almost mystical device; simple, yet with the regularity of a beating heart. It is a device that is capable of sensing and using, for its seemingly endless motion the invisible gravitational field that permeates us and our world, and which fixes us to the surface of this planet.

In 1668, the newly created Royal Society of London published *An Essay Towards a Real Character and a Philosophical Language* by John Wilkins (1614–72), the Bishop of Chester. Wilkins' essay included a short description of a system of weights and measures based upon a single 'universal measure' that could be used to define length, weight, volume, and money.

John Wilkins suggested a decimal system of measurement, with a universal standard of length derived through the use of a seconds' pendulum. This standard length could then be used to define area, volume, and weight using a well-defined volume of distilled rainwater. Wilkins' *Essay* is the first description of a complete system of measurement intended to be used by all nations. Indeed, Wilkins' proposal contained almost all of the essential elements of the Metric System of 1795, which could quite reasonably therefore be said to have originated in England in the 17th Century, and not in France during the late-18th Century. Following John Wilkins' first description of a possible international system of measurement, the development of the decimal metric system of measurements was inevitable even though Wilkins himself was not confident of its success. Wilkins wrote about his plans for a universal measure, 'I mention these particulars, not out of any hope or expectation that the World will ever make use of them, but only to show the possibility of reducing all Measures to one determined certainty.'

Following the publication of Wilkins' essay, *savants* in several countries took up and promoted his idea of an international system of units or weights and measures based on a single natural dimension. In 1670, Gabriel Mouton (1618–94), a French cleric and astronomer, promoted a system of measurement based upon the physical dimensions of the Earth, rather than a measurement based on the length of a seconds' pendulum.

Gabriel Mouton assumed that the Earth was a perfect sphere, and so a section along a meridian would be a circle. Mouton proposed that this 'great circle' should be divided into ever smaller angles and that these angles could be used to define a system of measurement. What was also proposed was that these division of these angles should be made using decimal arithmetic; that is, division by ten, rather than the old Babylonian sub-divisions of an angle based on arithmetic to the base 60 (an angle being divided into 60 min and each minute into 60 s). Mouton suggested that a minute of arc along the meridian be measured and defined as a unit of distance called a *milliare*. The Abbé Mouton also suggested dividing the *milliare* into *centuria*, *decuria*, *virga*, *virgula*, *decima*, *centesima*, and *millesima* by successively dividing by factors of ten. Mouton's *milliare* is the definition of a modern Nautical Mile (that is, one minute of arc of latitude along any meridian), which given the importance of maritime trade accounts for the longevity of this system of measurement.

This idea of establishing a universal system of units or a universal measure based on the dimensions of the Earth was taken up by many natural philosophers; it identified the need for precise measurements of our planet. One of the first surveyors who undertook the task of precisely determining the curvature of the Earth was the founder of a dynasty of French astronomers, Jacques Cassini (1677–1756). With his son, César-François (1714–84), Jacques Cassini surveyed a portion of the Arc of the Meridian from Dunkerque (Dunkirk) in France to Barcelona in Spain. This particular line passes through the Paris Observatory and is called the Paris Meridian, and it would be surveyed many times over the following century. These repeated measurements of ever-increasing precision yielded the first standard metre, which is still preserved in Paris. The surveying would be completed by Jacques' grandson Jean-Dominique (1748–1845)[3].

These early surveyors used a seconds' pendulum to define a length standard at a location, and in so doing quantified the local variations of gravity. Between 1739 and 1740, the astronomer Nicolas Louis de Lacaille (1713–62) together with Jacques Cassini again measured the Dunkirk–Barcelona Meridian. Their objective was to extend the earlier measurements, and their surveying formed the basis of the 'provisional metre' established in 1793. The northern and southern ends of the surveyed meridian are the Belfry in the centre of Dunkirk and the fortress of Montjuïc, respectively. Apart from defining the dimension of the universal measure or metre, these early surveyors refined the value of the Earth's radius and established that the shape of the Earth is oblate or slightly flattened near the North and South Poles, which had been predicted by Isaac Newton.

As the 18th Century drew to its close, two political events occurred which would have a profound influence upon the nature of the various systems of weights and measures still used throughout the world today. In North America, the British colonists rebelled and fought for their independence. The rebellion was successful, and by 1791 a new nation was born which was using the system of weights and measures they had inherited from England (see section 9.3). But would they wish to continue using this system, which tied them to the 'old country'? In Europe, the major political event of this period was the bankruptcy of France, and the collapse of the nation into the French Revolution of 1789.

2.4 'Dear boy …'

The reform of Julius Caesar's solar calendar proposed by Pope Gregory XIII in 1582 was rapidly adopted by the Catholic countries of Europe. There was, however, a great deal of reluctance from the new Protestant nations to be seen accepting anything suggested by the Pope. This, unforeseen consequence of the Reformation led to a Europe not so much of time zones, with an hour difference between two neighbouring nations, but a Europe where you could move from one month to the next, or vice versa, by crossing a frontier. By the late-16th Century, the solar calendar was out of step with Nature by more than a week, and more than a week had to be 'deleted' from the calendar to bring some kind of order to the slippage between man-made calendars and the seasons. Thus, the world crystallised into a mosaic of very different time zones. The Scandinavians did not accept the Gregorian

---

[3] The Paris Meridian is not our present line of zero latitude. The modern Prime Meridian runs through the Greenwich Royal Observatory, and so it is at Greenwich and not Paris that the world is bisected. This move to a Greenwich-based view of the world was decided by the International Meridian Conference, 1884, Washington DC. Modern satellite measurements of the geographical coordinate of Paris show how much the meridian line had moved; the Paris Observatory is at 48°50′0″N 2°21′14.025″E, the Greenwich Royal Observatory is at 51°28′40.12″N 0°00′0.5.31″W. The French did not accept the 1884 change until after 1945.

Calendar until the early-1700s, the British and the British Colonies (including the American Colonies) adopted the change in 1752 and the Russians did not accept the new calendar until the revolution of 1917, which meant that when the USA purchased Alaska from Russia in 1867, the local population suddenly found themselves, at the stroke of a pen propelled almost two weeks into the future.

The problem with accepting the reformed calendar, apart from old-fashioned sectarianism, was that quite a few days had to be deleted to bring a nation into coherence with those countries which had already accepted the change. Ordinary people thought that they were losing a part of their lives. And of course, unscrupulous politicians could readily fan the flames of religious hatred and ignorance, by playing on this fear of having your life shortened by an act of parliament. It was Philip Dormer Stanhope, Fourth Earl of Chesterfield (1694–1773) who as a member of the House of Lords introduced, and succeeded in passing a Bill for the reform of the Julian Calendar in Great Britain. The Chesterfield Act was passed in 1751, nearly two centuries after the reform proposed by Pope Gregory XIII.

Philip Stanhope was one of the great aristocrats of the 18th Century. He is best known today as the author of hundreds of letters, which he wrote to his illegitimate sons; in particular, to Philip Stanhope over a period of almost 30 years. These letters were intended to supplement the aristocratic education of the young man. And the Earl wasted no time in telling his sons how to behave, and most importantly for a young gentleman of no fortune, how to rise in an aristocratic society. The letters are certainly cynical and worldly, but also full of frankness, honesty, and wisdom. Indeed, the Earl's letters are so full of worldly wisdom that they even managed to shock our late-18th Century forebears. However, it was probably an exaggeration on the part of Dr Johnson to say that the letters 'teach the morals of a whore, and the manners of a dancing master'.

Of particular interest to us in a history of the world's measurement system is a letter written by the Earl to his son in 1751, just after the passage of the Chesterfield Act. The Earl's letter of 18 March 1751 O.S.[4] explains not only how the Earl managed to get Britain to finally accept the reformed Gregorian Calendar, but in the process demonstrates how one may successfully undertake a difficult project in science communication. In effect, the adoption of the Chesterfield Act required 11 days to be deleted from the calendar of 1752. The Earl writes, 'It was notorious that the Julian Calendar was erroneous, and had overcharged the solar year with eleven days .... It was not in my opinion, very honourable for England to remain in a gross and avowed error...; the inconvenience of it was likewise felt by all who had foreign correspondence, whether political or mercantile' [3].

'I determined therefore to attempt the reformation; I consulted the best lawyers and the most skilful astronomers, and we cooked up a Bill for that purpose [*adopting the Gregorian Calendar*].' But as with most fundamental changes in science and technology, which will affect the population in general, there was the problem of

---

[4] O.S. is the abbreviation for Old Style, and N.S. stands for New Style. The styles referring to the use of the reformed Gregorian Calendar. The difference between O.S. and N.S. would be the number of days deleted from the year by the reform.

communication; why was it so important to pass this Bill and effect the changes to the calendar?

'But then my difficulty began.' The Earl points out to his son how he thought it necessary for the House of Lords to think that he actually knew something about astronomy, and for him to make their Lordships believe that they also knew something of it themselves '... which they do not.' As the Earl says to his son, 'I could just as soon have talked Celtic or Sclavonian to them [*their Lordships*] as astronomy, and they would have understood me full as well ....'

So, the Earl decided to 'please rather than inform them [*the House of Lords*]'. 'I gave them therefore only an historical account of calendars ... amusing them now and then with little episodes; but I was particularly attentive to the choice of my words, to the harmony and roundness of my periods, to my elocution, to my action. This succeeded, and will ever succeed; they thought I informed, because I pleased them; and many of them said, that I had made the whole very clear to them, when God knows, I had not even attempted it.' More than two hundred and fifty years after the Earl wrote these disarmingly honest comments about the nature of political institutions, one can only smile—little has changed.

The Earl's aristocratic, disdainful manner in treating with his supposed social equals (the other Peers of the Realm) previews the manner in which that other great aristocrat who involved himself in fundamental changes to a nation's system of weights and measures, Charles Maurice Talleyrand-Périgord, the Bishop of Autun, and who went on to become Prince Talleyrand-Périgord, persuaded the National Constituent Assembly in 1790 to set about a total reform of the systems of weights and measures then in use in France (see Preface). Both the Earl of Chesterfield and the Prince de Talleyrand-Périgord dazzled, impressed and convinced their respective audiences that they knew something about the subject under discussion, and so they achieved what they wanted. As the Earl of Chesterfield told his son, 'lay aside all thoughts of all that dull fellows call solid, and exert your utmost care to acquire what people of fashion call shining. Prenez l'éclat et le brillant d'un galant homme'. Whatever Dr Johnson might have said about such sound fatherly advice, it is certainly true to say that the best way of instructing a dull audience in something they should know, but in which they have absolutely no interest, or understanding is by seducing them. As the Earl puts it so succinctly, 'All your Greek will never advance you ...; but your address, your manner, your air, if good, very probably will.'

In the late-1780s, French weights and measures were in total confusion. At a time when the French economy was beginning to industrialize in response to similar developments in England, which had already created a national system of weights and measures, France was unable to compete as it did not have any means of standardizing manufacture. In England, factories could out-produce manual fabrication, because there was uniformity of measurement, and standardization of production.

The American colonies sent ambassadors to France, which was actively helping them in their struggle against Britain. Thomas Jefferson (1743–1826), who became the third president of the USA, served as Ambassador to France where he was in regular contact with British and French *savants* as they formed their ideas about new, more natural systems of measurement. And it was from this exchange of ideas that the political leaders in America began to ponder the appropriate system of weights and measures for their young nation, which would ensure that they were able to compete effectively, and independently on an international stage.

During his period of office as Ambassador to France, Thomas Jefferson visited London in 1789. While in London, the political situation in France deteriorated, and to avert bankruptcy King Louis XVI convened the *États Généraux* (Estates General) to impose new taxes. The *États Généraux* was a meeting of the three estates, or groups who were seen as constituting the nation: the First Estate was the clergy, the Second Estate was the nobility and the Third Estate, the bourgeoisie; the urban workers and the peasants were rather left out of things.

In March and April of 1789, King Louis XVI ordered that a list of grievances (*cahiers de doléances*) be drawn up for discussion by the *États Généraux*. Issues related to weights and measures were a constant theme in these lists of grievances, especially about how they related to rents, tithes and taxes. Here are some examples from the *cahiers de doléances* of 1789:

- Are the seigneurs [the nobles] not obliged to present to us some documents in support of the size of the measures they employ?
- In cases where the seigneurs' measures are found to be in excess of what they should be, are they not obliged to restore to their rent payers the resultant excess amounts?
- The seigneurs' agents are employing two different measures: one of them is larger than the proper one and is used in the collection at the granary. The payers are not blind to this injustice, but are too afraid to protest.
- The seigneurs and their agents use falsified and oversized measures.

The *États Généraux* had not met since 1614. Although there had been previous attempts to impose the units of measurement habitually used in Paris upon the whole of France, such moves had always been opposed by the church, the guilds of artisans and professionals, and the nobles who all benefited greatly from confusion in measurements.

No doubt the king hoped that by giving the people of France the opportunity of expressing their hopes and grievances directly to himself, this would help to relieve political tension. However, writing the *cahiers de doléances* forced the people to think carefully about their other grievances, and so added greatly to an air of revolutionary expectation in France.

Further reading

[1] Nelson R and Olsson M G 1987 The pendulum—Rich physics from a simple system *Am. J. Phys.* **54** 112–21
[2] Sobel D 2005 *Longitude: the True Story of a Lone Genius who Solved the Greatest Scientific Problem of his Time* (New York: Walker and Co.) 10th anniversary edition
[3] Stanhope (Fourth Earl of Chesterfield) P 1986 *Letters to his Son and Others (Everyman Classics)* (London: Random House)

Defining and Measuring Nature (Second Edition)
The make of all things
Jeffrey Huw Williams

Chapter 3

Measurement in the modern world (I)

3.1 Surveying and measuring the Earth

Before looking at one of the triumphs of Enlightenment science, the metric survey of the 1790s, let us first briefly consider some earlier surveys of the Earth's surface. We begin with the person who made the first attempt to define our planet, that is, to determine its circumference, Eratosthenes of Cyrene (c. 276 BCE–c. 195/194 BCE).

Eratosthenes was a Greek polymath, famous in his own day for the study of: mathematics, geography (he invented the subject), poetry, astronomy, and music theory. He was born in, what is modern Libya, and became the third chief-librarian of the Library of Alexandria; the storehouse of ancient learning founded by Alexander the Great, and destroyed by the legionaries of Julius Caesar in 48 BCE. Today, Eratosthenes is best known for being the first person to calculate the circumference of the Earth, which he did by comparing the angles of the shadows cast by the mid-day Sun at two places, a known north–south distance apart. His measurement was remarkably accurate. He was also the first to calculate the tilt of the Earth's axis, which gives us our seasons; again, with remarkable accuracy. Indeed, Eratosthenes created the first map of the world, incorporating parallels and meridians based on the available geographic information.

3.2 The circumference of the Earth

Eratosthenes calculated the Earth's circumference from two measurements of the length of a shadow cast at the same time in two widely separated locations. He knew that at local noon on the Summer solstice in Syene (modern Aswan, Egypt), the Sun was directly overhead, because the shadow of someone looking down a deep well at that time in Syene blocked the reflection of the Sun on the water. He then measured the Sun's angle of elevation at noon, on the same day in Alexandria by using a

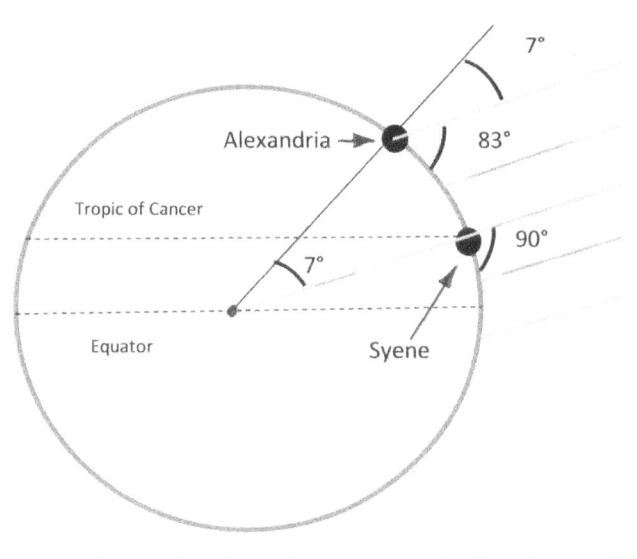

Figure 3.1. The geometry of Eratosthenes' experiment. We see the incoming parallel rays of sunlight. To determine the circumference of the sphere, one must know the distance between Alexandria and Syene, and the angle made by the shadow of a gnomon of known height at Alexandria.

vertical rod, known as a gnomon[1], and measuring the length of its shadow on the ground. Using the length of the gnomon, and the length of its shadow as the sides of a triangle he calculated the angle of the Sun's rays (see figure 3.1). This turned out to be about 7°. Taking the Earth as spherical, and knowing both the distance and direction of Syene from Alexandria, he was able to calculate the Earth's circumference.

Eratosthenes would have required a detailed measurement of the distance between Alexandria, in the Nile Delta to Syene, in the south of Egypt. This information was available from the work of the annual surveying trips conducted by map-makers of the Pharaohs of Ancient Egypt. Pharaonic priests (they were also responsible for metrological standards, see chapter 1) gave a distance between Syene and Alexandria of 5000 stadia[2]. Some claim Eratosthenes used the Olympic stadion of 176.4 m, which gave the Earth a circumference of 44 100 km, an error of about 10% when compared to the modern value. As with all surveyors, Eratosthenes needed to make assumptions to be able to derive any result from his simple experiments.

---

[1] The projecting piece on a sundial that casts a shadow indicating the time.

[2] The stadion was an ancient Greek unit of length, based on the circumference of a sports stadium of the time. According to Herodotus, one stadion was equal to 600 Greek feet. However, the length of the foot varied in different parts of the Greek world, and the length of the stadion has been the subject of argument and hypothesis for centuries.

3.3 The Chinese survey

As we will see later, by the 1790s the European Enlightenment had come up with the idea of constructing a new decimal metrology based on a single measurement of length. Such ideas, however, have a long history, and it is to Ancient China that we must turn for the first consistent use of a coherent system of weights and measures; particularly, in the decrees of Qin Shi Huang, who became First Emperor in 221 BCE.

Given the size of China, it is perhaps not surprising that an early effort was made to fix terrestrial length measurements in terms of astronomical observations. The harmony of the coupled celestial and terrestrial worlds has always been a characteristic of Chinese thought, reflecting the astral character of the ancient Chinese state religion. It was an early idea of Chinese *savants*, going back before the time of Confucius (551−479 BCE), that the shadow-length of a standard height (an eight-foot gnomon), at the Summer solstice increased by one inch for every thousand *li* (equivalent to 1500 *chi* or Chinese feet) north of the Earth's 'centre' (*Yang chhêng*, the Tropic of Cancer), and decreased by the same proportion as one went south. This rule of thumb remained current until the Han Dynasty (205 BCE−220), when detailed surveying of the expanding Chinese Empire showed it to be incorrect. But it was not until the Tang Dynasty (618−907) that a systematic effort was made to determine a range of latitudes. This extensive Tang survey had the objective of correlating the lengths of terrestrial and celestial measures by finding the number of *li* that corresponded to 1° of polar altitude; that is, terrestrial latitude; thereby fixing the length of the *li* in terms of the Earth's circumference. This Chinese meridian survey takes its place in history after Eratosthenes, but more than a thousand years before the metric survey of the 1790s.

It was the Tang emperor, Xuanzong (685–762), who commanded the clerks of the Bureau of Astronomy to determine the size of his empire. The majority of the Chinese surveying measurements were undertaken between 723 and 726 by the Astronomer-Royal, Nankung Yüeh, and his assistant I-Hsing (683–727), a Buddhist monk. Eleven stations were established for the measurements (including *Yang chhêng*), with polar altitudes ranging from 17.4° at *Lin-i* near modern Hué in Annam (Vietnam) to 40° at *Weichow*, near modern *Ling-chhiu*, near the Great Wall in northern Shansi (a similar latitude to Beijing). These stations were not on a single north–south axis and were not uniformly distributed north–south; only one station was near the northern border of China, and two stations were in the far south. This asymmetry in the distribution of stations would seriously limit the ultimate precision of the survey. The length of the distance investigated by the Tang surveyors was 7973 *li*, which is about 2500 km; the French metric survey was from Dunkirk to Barcelona, a distance of about 1000 km.

The method used by the Tang surveyors was the same as that used by the French metric surveyors a thousand years later, that is, trigonometry. We read in the Tang chronicle (*Thang Hui Yao*) [1] '… Then on the basis of the northern and southern sun-shadows I-Hsing made comparisons and estimates, using the right-angle triangle method to calculate them …' [Needham *Science and Civilization* p 46]. This is precisely what Delambre and Méchain did in France in the 1790s, but to a much higher precision; that is, with fewer approximations. The main result of this Chinese field-work was that the difference in shadow-length, for an eight-foot gnomon was found to be close to four inches for each 1000 *li* north and south, and that the terrestrial distance corresponding to 1° of polar altitude was calculated to be 351 *li* and 80 *bu* (1 *li* = 360 *bu*).

This Chinese survey is today practically unknown outside of China, yet it represents an outstanding achievement given the spaciousness and amplitude of its plan and organization, and one of the earliest uses of advanced mathematics needed to compute the final result. These results were known in 18th Century Europe, as they were commented upon by Leonard Euler and later by Pierre Simon de Laplace. However, while the French metric survey obtained a routine precision of a few parts in 10^5 in distance, the earlier Chinese survey could boast only of a precision of 1 part in 10^3.

3.4 La Révolution Française

The systems of weights and measures that existed in France on the eve of the revolution were complex and had evolved to institutionalize fraud. The difficult lives of the French people were further burdened by more than 700 differently named measures, and untold units of the same name but of differing size: in Paris, the *pinte* (a unit of volume) was equivalent to 0.93 l, but in nearby Saint Denis it was 1.46 l, and in Seine-en-Montagne it was 1.99 l. These non-uniform measures were enforced in law, as well as by custom. A landlord might wish to have his rent paid in bushels of grain or hogsheads of beer in the largest measures in use in the neighbourhood, but he would then always prefer to sell according to the smallest unit.

On 4 August 1789, the National Constituent Assembly, which had grown out of the *États Généraux*, abolished the feudal system of land management. These changes meant that the nobles no longer had a monopoly on weights and measures. However, this rapid suppression of the existing system of weights and measures effectively meant that France had no measurement system at all. And, not surprisingly, this hasty action before a new system of weights and measures had been implemented, precipitated further confusion.

This metrological void had somehow to be filled, and filled quickly against a background where powerful social groups were attempting to safeguard their vested interests. At the same time as the National Constituent Assembly was seeking to fulfil the dreams of progressive thinkers, they were simultaneously trying to satisfy the aspirations of the more complacent and non-radical *bourgeoisie*. It was at this moment that, perhaps the most fascinating, colourful and least-expected person to appear in a history of metrology, Charles Maurice de Talleyrand-Périgord (1754–1838), Bishop of Autun, but usually known simply as Talleyrand, was ready and waiting.

In 1790, Talleyrand made a report to the National Constituent Assembly on the state of weights and measures then in use in France; he suggested a fundamental reform of the whole system (see Preface). Eventually, the deputies decided that there should be a radically new system of weights and measures, based on a new fundamental measure of length (John Wilkins' universal measure) determined by the length of the seconds' pendulum at Paris. Talleyrand further suggested that the *Académie des sciences* collaborate with the Royal Society of London in defining this new fundamental unit. The National Constituent Assembly and the now constitutional King Louis XVI approved Talleyrand's proposal, but sadly nothing came of this unique opportunity for international, scientific collaboration.

After having been persuaded by Talleyrand, the National Constituent Assembly decreed that all measurement standards in France should be sent to the *Académie des sciences* in Paris who would then issue new standard measures. However, neither Talleyrand nor the National Constituent Assembly had much of an idea of the magnitude of this task. The politicians thought that new standards could be adopted and copies of these new standards could be distributed all within six months. In this, they were very much mistaken.

Events were also moving rapidly on the other side of the Atlantic. The Americans had gained their independence, and on 8 January 1790 George Washington sent his first message to the new US Congress, where he reminded the legislators of their responsibility for weights and measures: 'A uniformity of weights and measures is among the important objects submitted to you by the Constitution, and, if it can be derived from a standard at once invariable and universal, it must be no less honorable to the public council than conducive to the public convenience .... Uniformity in the currency, weights, and measures of the United States is an object of great importance, and will, I am persuaded, be duly attended to.'

The legislators had more important things on their minds, as Washington was obliged to repeat his call for urgent attention to be given to the uniformity in the currency, weights and measures for the new United States in his second and again in his third annual presidential messages to Congress. One can see from Washington's message ('derived from a standard at once invariable and universal') that he was referring to a system of weights and measures derived from John Wilkins' universal measure.

Eventually, the Congress responded to the President's request by asking Secretary of State Thomas Jefferson to prepare a report on the subject of weights and measures in the new nation. This was a subject upon which Jefferson was eminently qualified to report, as he had familiarized himself with all the major European ideas on this subject during his time as American Ambassador to France.

Meanwhile, back in Europe, knowing that Sir John Riggs Miller (1744–98), the Member of Parliament for the 'rotten borough' of Newport in Cornwall, had raised the question of an improved system of weights and measures (one based on scientific principles) in the House of Commons in 1789, Talleyrand wrote to him in his capacity as a deputy of the clergy in the National Constituent Assembly on 29 March 1790 (this is Sir John Riggs Miller's translation of Talleyrand's letter).

Sir, I understand that you have submitted for the consideration of the British Parliament, a valuable plan for the equalization of measures: I have felt it my duty to make a like proposition to our National Assembly. It appears to me worthy of the present epoch that the two Nations should unite in their endeavour to establish an invariable measure and that they should address themselves to Nature for this important discovery ...

... Too long have Great Britain and France been at variance with each other, for empty honour or for guilty interests. It is time that two free Nations should unite their exertions for the promotion of a discovery that must be useful to mankind.

Talleyrand's idea was to avoid nationalism in trade and metrology by the creation of a new, standard of measurement derived from Nature (*pris dans le nature*) and therefore, supposedly, acceptable to all nations. Talleyrand further suggested that the National Constituent Assembly, the British Parliament, the *Académie des sciences* of Paris, and the Royal Society of London should undertake preliminary joint work on this project. He wrote 'Perhaps this scientific collaboration for an important purpose will pave the way for political collaboration between the two nations.'

Of course, there were some problematic details to perturb the collaboration. The measurement of length being proposed for the determination of the new universal measure would have been made using a seconds' pendulum. And unfortunately, the precise value of this new universal unit of length would have depended upon the local value of gravity at the site where the measurements were made. Thus, the fundamental or universal unit of length, upon which the entire system of weights and measures would have been based, would have been tied to the value of gravity in Paris, where Talleyrand proposed making the measurement, or London where Riggs Miller proposed making the measurement, or Monticello in Virginia where Jefferson proposed making the measurement.

Nationalism is invariably present when you scratch the surface of any international scientific endeavour. However, this possible tri-partite scientific endeavour must surely rank as one of the great missed opportunities of history.

In April 1790, Riggs Miller informed the British Parliament of Talleyrand's letter and spoke in favour of the scheme, as the reform of international metrology would confound those who were unscrupulous in trade; as Riggs Miller commented, the current systems of measurement leads to '... the perplexing of all dealings, and the benefiting of knaves and cheats'.

In May 1790, Talleyrand submitted a proposal to the National Constituent Assembly for a decimal system of stable, unvarying and simple units of measurement. These were to be based on the length of the seconds' pendulum at Paris. At Talleyrand's suggestion, the French National Constituent Assembly adopted this new system of measurement (see Preface).

In July 1790, Secretary of State Thomas Jefferson submitted his report *Plan for Establishing Uniformity in the Coinage, Weights, and Measures of the United States* to the US Congress. However, no official action was taken and Congress passed no legislation relating to weights and measures. Not surprisingly, Jefferson's report referred to many of the scientific investigations being undertaken to reform the French system of weights and measures.

In August 1790, King Louis XVI authorized the scientific investigations, which would ultimately lead to the creation of the Metric System in 1795. The *Académie des sciences* was given responsibility for the research, and they passed the request to a recently appointed commission to do the detailed scientific and technical work and to produce a report for the National Constituent Assembly. The members of this Commission on Weights and Measures at the *Académie* reads like a 'who's who' of 18th Century science. In addition to the mathematician and philosopher Condorcet (Marie Jean Antoine Nicolas de Caritat, *marquis* de Condorcet 1743–94) who was secretary, the initial members were: the mathematician and naval officer Jean-Charles, *chevalier* de Borda (1733–99); the physicist and mathematician Charles-Augustin de Coulomb (1736–1806); the Italian-born astronomer and mathematician Joseph Louis de Lagrange (1736–1813); the mathematician and astronomer Pierre-Simon, *marquis* de Laplace (1749–1827); the chemist Antoine-Laurent de Lavoisier (1743–94); and the botanist and metallurgist Mathieu Tillet (1714–91) who had been responsible for the fabrication of previous measurement standards.

The Commission's report was presented to the National Assembly in October 1790, and it recommended the decimal division of money, weights, and measures. In detail, they recommended 'that the length of a meridian from the North Pole to the Equator be determined, [*and*] that 1/10 000 000th of this distance be termed the metre [*this was then to be the universal measure*] and form the basis of a new decimal linear system, and, further, that a new unit of weight should be derived from the weight of a cubic metre of water [*that is, the modern metric tonne or 1000 kilograms*]'.

The commission specifically rejected the use of the seconds' pendulum to define the new unit of length, because the oscillations of the pendulum could not be considered truly harmonic, and could not be modelled to derive a value of the new unit of length (the metre) which would be sufficiently accurate. This report also recommended that the new system of weights and measures should be decimal and that it includes a list of prefixes for decimal multiples and sub-multiples. The system proposed was to become the Metric System, which as the International System of Units is now in use in every nation. At the same time, the *Académie des sciences* also recommended the decimal division of the new money to be issued by France. The previous unit of currency had been based on the *livre* or pound, and had the same non-decimal sub-divisions as did the currency of Great Britain until 1971.

By the end of 1790, the *Académie des sciences* had appointed an enlarged scientific commission to consider the whole question of weights and measures, and how best to achieve the recommendations in the report of October. The members of this enlarged scientific commission were: Condorcet, Coulomb, Lavoisier, Lagrange, Laplace, de Borda, Gaspard Monge, *comte* de Péluse (1746–1818) a mathematician and the founder of the *Ecole Polytechnique*, Jean-Dominique Cassini, *compte* de Cassini (1748–1845) the astronomer and surveyor who had already been involved with his father in surveying the meridian from Dunkirk to Barcelona, another astronomer and surveyor Pierre-François-André Méchain (1744–1804), the mathematician Adrien-Marie Legendre (1752–1833), the astronomer and surveyor Jean-Baptiste-Joseph Delambre, *chevalier* Delambre (1749–1822), and General Jean

Baptise Marie Charles Meusnier de la Place (1754–93) who did not participate in the scientific work of the commission as he died of wounds received at the siege of Leipzig [2].

3.5 Defining the size of the world

The science commission of the *Académie des sciences* had recommended a measurement of the new standard of length, the metre, based on a detailed survey along the meridian extending from Dunkirk to Barcelona, which had already been surveyed and measured by de Lacaille and César-Francois Cassini in 1739. The commission calculated that if they could measure a significant piece of the meridian, the rest could be estimated. Both ends of the line to be surveyed needed to be at sea level, and as near to the middle of the Pole-to-Equator Quadrant as possible to minimize errors. Fortunately for them, the only appropriate meridian is about a tenth of the distance (about one thousand kilometres) from the Pole to the Equator and it runs through Dunkirk and Barcelona, so most of the distance to be surveyed lay conveniently inside France, a fact that did not escape the more nationalistic attention of observers such as Thomas Jefferson.

Condorcet appreciated the potential for such nationalist views when he wrote 'The Academy has done its best to exclude all arbitrary considerations—indeed, all that might have aroused the suspicion of its having advanced the particular interests of France; in a word, it sought to prepare such a plan that, were its principles alone to come down to posterity, no one could guess the country of its origin.' The Legislative Assembly endorsed the proposal from the *Académie des sciences* and directed that the detailed survey be made as soon as possible, and enacted the necessary legislation on the 26 March 1791.

Although the *Académie des sciences* finally chose that the metre would be exactly a ten millionth of the distance between the North Pole and the Equator, their choice also defined this distance as being precisely 10 000 000 m. Unfortunately, an error was made in the commission's initial estimation, because the wrong value was used in correcting for our planet's oblateness. We now know that this Quadrant of the Earth is actually 10 018 750 m. One should never forget that these *savants* were not only setting out to create what they saw as a new fundamental system of units based on the dimensions of the Earth, but they were also imposing models and views about the character of the Earth. In 1791, a handful of mathematicians, guided by the writings of Isaac Newton, imposed a definite shape and size to our planet; the Earth shrank, and became precisely known.

3.6 The metric survey

On 19 June 1791, Jean-Dominique de Cassini (head of the Royal Observatory) had secured an audience with King Louis XVI for some of the members of the commission of the *Académie des sciences*. At six in the evening, Cassini, Legendre, Méchain, and Borda (the inventor of the repeating circle which was hoped would increase the level of precision possible in surveying and thus allow the determination of the metre) presented themselves at the *Palace du Tuileries*. A small group of eminent astronomers and mathematicians had come to convince the now constitutional monarch, King Louis XVI that the metre was something worth achieving.

History has not been kind to Louis XVI. He has a reputation for having been naïve and something of a simpleton, but he had hidden talents. The king was a skilled instrument (watch) maker and something of a cartographer. The king also took a close interest in the cost and necessity of the proposed survey. Turning to the head of the Royal Observatory he asked, 'How's that, Monsieur Cassini? Will you again measure the meridian your father and grandfather measured before you? Do you think you can do better than they?'

Monsieur Cassini (the third generation of a dynasty of directors of the Royal Observatory) was not unused to conversing with the monarch. 'Sire, I would not flatter myself to think that I could surpass them had I not a distinct advantage. My father and grandfather's instruments could but measure to within fifteen seconds [*of a minute of a degree*]; the instrument of Monsieur Borda here can measure to within one second.' (The details of these conversations were published in the *Comptes rendus* of the *Académie des Sciences*.) Figure 3.2 is a photograph of a modern reconstruction of Borda's surveying instrument (Borda's circle).

King Louis XVI quickly gave his formal approval for the new survey of the Dunkirk–Barcelona Meridian. Then, in the early hours of the next day, the king and his family attempted to escape from France (the 'Flight to Varennes'), but they were arrested, returned to Paris and then imprisoned. What must the king have been thinking when he was contemplating his escape and the suppression of

Figure 3.2. Borda circle, on display at *Conservatoire national des arts et métiers*, Paris. The height of the surveying instrument, which is made of brass, is about the height of an adult. The angular gradations on the large brass circle were decimal, not to the base 60. (Image from https://en.wikipedia.org/wiki/Repeating_circle#/media/File:Repeating_circle-CnAM_1842-IMG_4998-gradient.jpg; it has been obtained by the author from the Wikimedia website where it was made available by Rama under a CC BY-SA 3.0 FR licence. It is included on that basis. It is attributed to Rama.)

the revolution, while trying to comprehend the mathematics of the *savants* and astronomers as they attempted to create a new universal system of weights and measures?

However, under arrest or not, Louis XVI was still the king, and from his prison cell he issued the proclamation that directed Jean-Baptiste Delambre and Pierre François-André Méchain, to undertake the surveying operations necessary to determine the length of the metre by precisely measuring the distance, along the meridian from Dunkirk to Barcelona. The king also issued orders to Baron Gaspard Clair François Marie-Riche de Prony to produce new trigonometry tables which would be needed to calculate the new universal measure from the surveying work of Delambre and Méchain.

The survey from Dunkirk to Barcelona was a major undertaking. Lavoisier called it 'the most important mission that any man has ever been charged with'; the measurements were designed to have been completed within a few months, yet it took the two surveyors from May 1792 to September 1798 to complete the work. The technical difficulties were compounded by more practical problems such as civil and international wars. France was in uproar with some cities restoring (with British aid) governments favourable to the monarchy, for example, Toulon. At one point, the surveyors were themselves accused of wishing to restore the monarchy, because the traditional white flags which they planted on the top of triangulation points on hills appeared to resemble too closely the old royal Bourbon flag of France; a white background with small gold *fleurs de lyses*.

The surveying method used by Delambre and Méchain was triangulation, where they had to accurately measure the angles in each of the triangles into which they had subdivided the territory to be surveyed (see figure 3.3, first part). The surveyors emphasized the decimal aspect of the Metric System by discarding the traditional (Babylonian system where angles are sub-divided into 60th parts) degrees and minutes of angular measurement, and instead divided the quadrant into one hundred grads that were then sub-divided into one thousand arc-minutes, that were in turn sub-subdivided decimally into arc-seconds.

At the completion of the measurements of each of the three angles in each of the more than 150 triangles into which the surveyors had divided the distance between Dunkirk and Barcelona (see figure 3.3), they had to measure as precisely as possible the length of one of the sides of one of these triangles. Then using trigonometry, they could calculate all of the distances, in all of the triangles. Consequently, two of the most important measurements were made by repeatedly laying a set of two-*toise* (a *toise* was a distance of about two yards) massive Platinum rulers end to end along two straight, flat roads.

In preparation for this final calculation of the metre, in September 1795 Méchain found a suitably flat stretch of road near Perpignan, the length of which he measured precisely with his Platinum *toise* measuring sticks. This flat stretch of road near the Spanish frontier was found from measurements made over several weeks by precisely placing the measuring sticks one in front of the other, to be a little over 6000 *toises*. This distance was one of the sides of one of the triangles being surveyed, and it would serve as a test for the quality of the angle measurements.

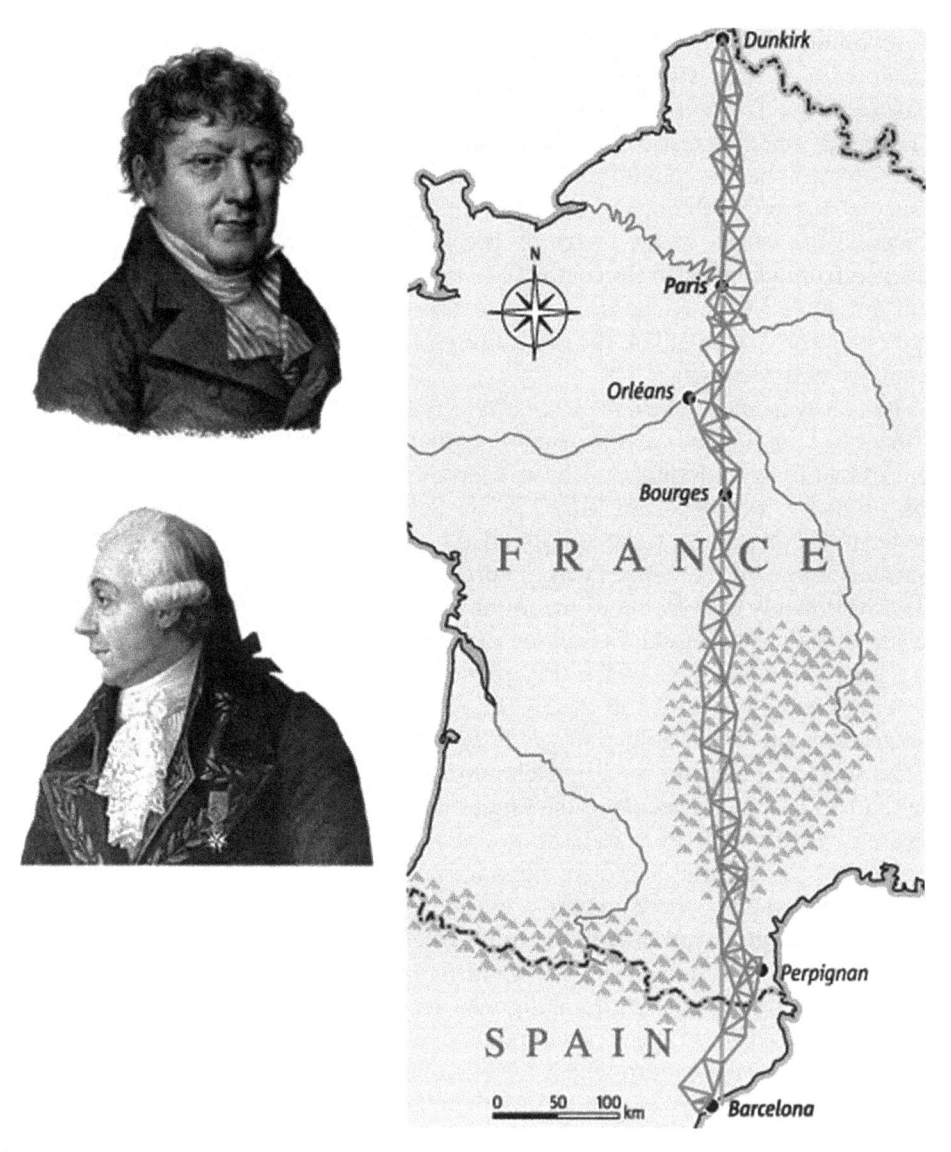

Figure 3.3. The route surveyed for the determination of the metre by Jean-Baptiste-Joseph Delambre (upper inset, reprinted from J-B-J Delambre 1912 *Grandeur et figure de la Terre*, published by G Bigourdan, Gauthier-Villars, Paris) and Pierre Méchain (lower inset). One can clearly see the more than 150 triangles into which the distance from Dunkirk to Barcelona had been divided for this triangulation exercise. Delambre worked south from Dunkirk, and Méchain worked north from Barcelona. (Image provided by kind permission of Mr Ben Gilliland, www.cosmonline.co.uk).

More detailed maps of the areas surveyed are given below.

This final test of the quality of the surveying and hence of the precision with which the metre had been determined, came in 1798 when Delambre took almost two months to measure the length of one of the sides of one of the surveyed triangles in the countryside near Paris. Again, the precise measurement of length was made with

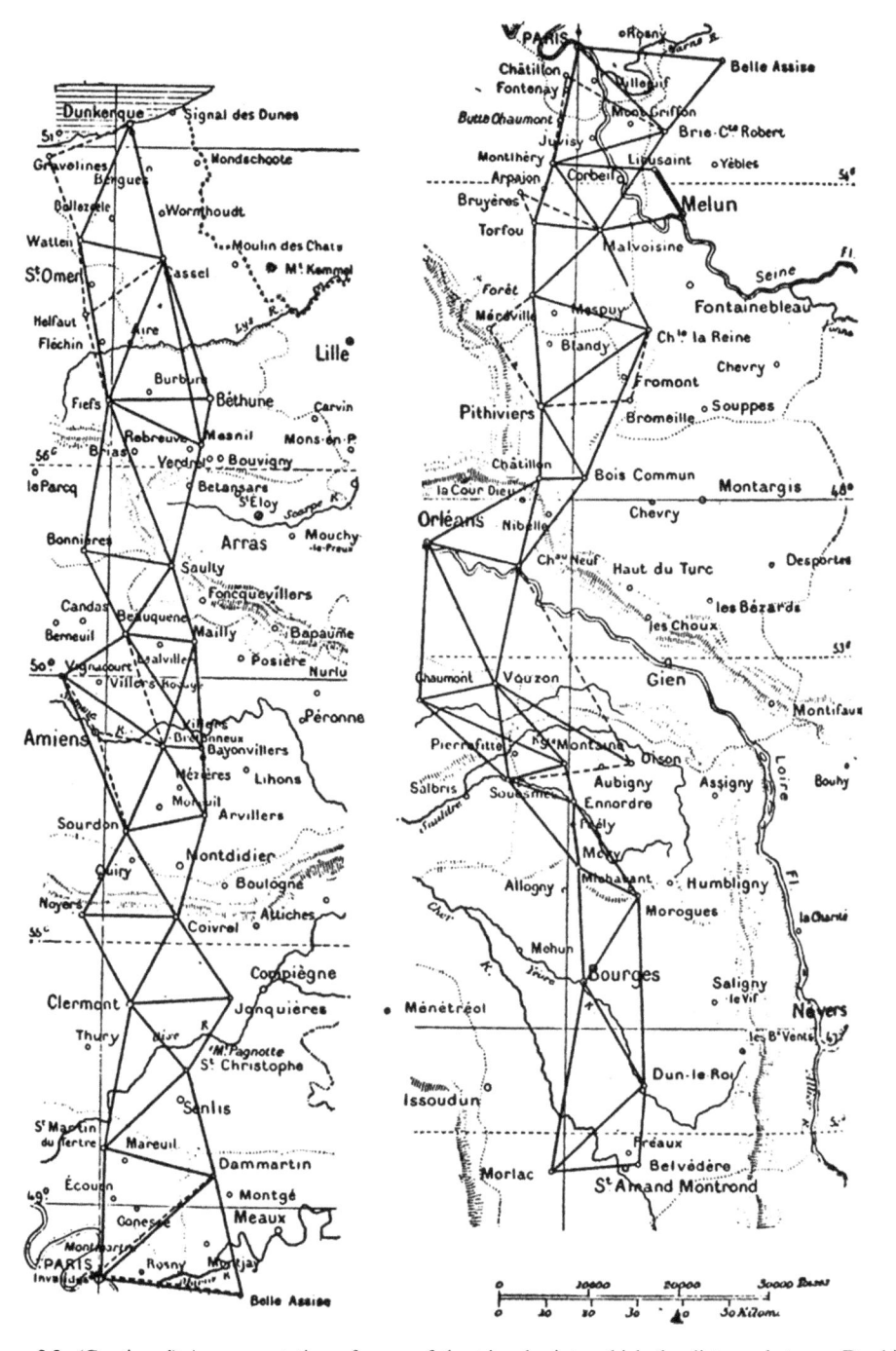

Figure 3.3. (Continued) A representation of some of the triangles into which the distance between Dunkirk and Barcelona was divided for the metric survey of 1792–98, and in which each of the angles had to be determined with the greatest precision. The present image shows a section of the surveying plan (from J-B-J Delambre *Grandeur et figure de la Terre*, published by G Bigourdan, Gauthier-Villars, Paris, 1912). The size of the triangles depended upon the availability of appropriate hills, mountains, church towers, etc. The units in the above schematic are kilometres and toises. (See text for details.)

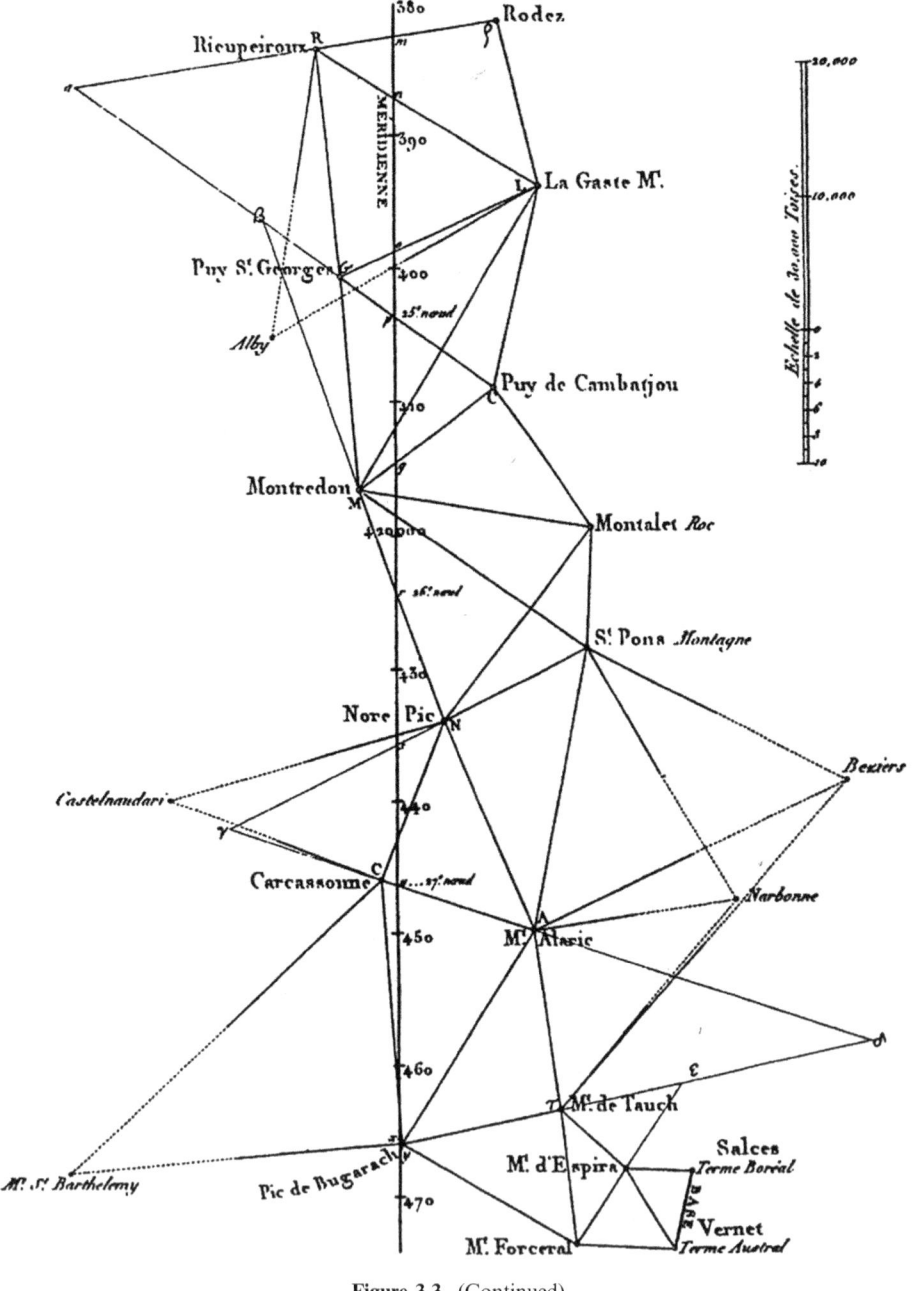

Figure 3.3. (Continued)

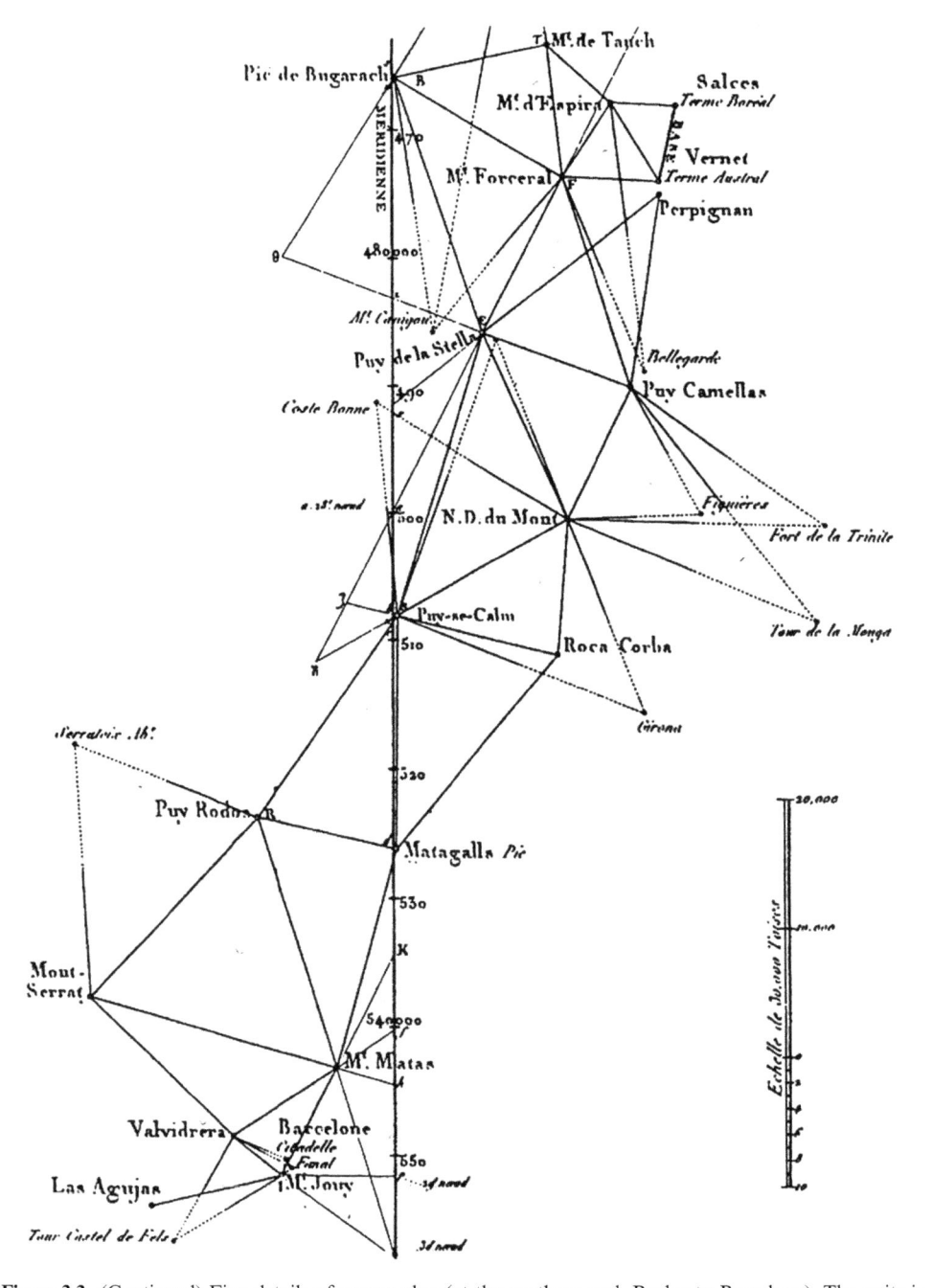

Figure 3.3. (Continued) Fine details of survey plan (at the southern end, Rodez to Barcelona). The units in these two detailed schematics are toises, a non-metric unit equivalent to about a fathom or two yards.

a pair of two-*toise* platinum measuring sticks. After this measurement of distance was completed in September 1798, the two surveyors met to compare their two length measurements. From the measured distance in the countryside near Paris, 6075.90 *toises*, they calculated a theoretical value for the length of the side of the triangle near Perpignan, 6006.198 *toise*. Méchain then revealed that he had actually measured this distance to be 6006.27 *toise*. As the surveyors commented 'the difference is negligible'. The years of surveying were over [3].

But even after almost seven years of toil, the astronomers' carefully calculated metre turned out to be no more precise than the preliminary estimate of the *Académie des sciences*, which the French authorities, to the annoyance of many *savants*, had promoted in the interim as a provisional unit. Sadly, because of the political situation in France and the physical and financial hardship of the surveying process, the two surveyors and their teams took a lot longer to complete their task than had originally been expected. As time went on, the politicians in Paris realized that to satisfy the demands of the people for a new system of weights and measures they could not wait for the precise measurements of Delambre and Méchain, so they adopted a provisional metre based upon the survey of the same Meridian carried out by César-Francois Cassini and then made and distributed wooden and steel 'provisional metre sticks'.

3.7 The error in all things

Many have commented on Méchain's later confession that his calculations contained a 'secret error'; a slightly divergent latitude reading, which Méchain identified in his own work but could not bring himself to admit (although his colleagues suspected that he was keeping quiet about it). These errors in the measurements of the position of some stars in the northern sky in Barcelona threw off other calculations and drove Méchain to the brink of suicide. Sadly for Méchain, statistics had yet to be invented[3], and so he could not use a trivial mathematical procedure such as least-squares analysis to weight the quality of his individual measurements; he was obliged to treat each measurement as being equally precise and accurate, and made under exactly the same conditions and for the same time period. This was not the case, but there was nothing that Méchain could do about it, and he had a nervous breakdown from not being able to admit that his averaged measurements were less precise than they should have been theoretically.

Today, we think nothing of making measurements of distance over extended sections of the globe. We have a network of satellites in orbit around the Earth, and this huge financial investment has made global positioning something that you have in your car. But what of the measurements of those two late-18th Century surveyors, Delambre and Méchain, who set out into a war zone with the very latest of surveying equipment and measured the distance from Dunkirk to Barcelona with the greatest of precision?

---

[3] A modern analysis of a large set of important, complex experimental data would likely involve a least-squares analysis. But this technique was not available until Gauss used it to correctly analyse the results of the metric survey, after they had been published in Germany in 1799.

The angle measuring repeating circles (see figure 3.2), which had been designed by Jean-Charles de Borda and used by the two surveyors to survey the meridian from Dunkirk to Barcelona where an enormous improvement over previous surveying instruments. In themselves, the circles were very stable by virtue of being massive and large; however, the problem was that they had usually to be mounted above ground level to make a measurement. And consequently, the wooden platforms that the precision repeating circles were mounted upon were the main source of the error in the measurement; particularly, when the weather was stormy and windy.

To gain an idea of the precision of the metric survey, consider some data measured by Delambre near Dunkirk. Here, he needed to establish the latitude of the observations, and to do this, he made detailed measurements of the transits of various well-known stars. Delambre made dozens of observations, which gave him a latitude of $51°\,2'16.66''$; a value which changed by a fraction of an arc-second when he removed what he perceived to be the least-reliable measurement. Let us therefore assume a precision of 0.1 arc-second; to relate this estimate of measurement uncertainty to a distance, consider the circumference of the perfectly spherical Earth as being 40 000 km. By dividing this circumference by 360, we determine the distance equivalent to measurement of a single degree of arc, then by dividing by 60 and then again by 60 we determine the distance equivalent to a single arc-second; which gives an uncertainty of about 4 m in the measurement of distance.

That is, these late-18th Century surveyors were capable of determining the location of an object on the surface of the Earth to a precision of about 14 feet, provided they had sufficient time. But this remains an amazing feat when one considers that a commercial GPS system is only accurate to about three times this level of precision; and the GPS represents an enormous investment in space-age technology[4].

The final results of the survey not only confirmed the value of the provisional metre established in 1793, but also produced something that was not anticipated; genuine new science. Delambre and Méchain had not set out to do basic research, but merely to improve the precision of something that had previously been measured. But what they discovered was that the Earth was even less spherical than Isaac Newton had calculated. According to surveying data determined by French *savants* in the mid-18th Century in Peru and France, the eccentricity of the Earth was roughly 1/300; that is, the Earth's radius to the Poles was 1/300 or 0.3% shorter than a radius to the Equator. Delambre and Méchain had observed that this eccentricity (from Dunkirk to Barcelona) was about 1/150, or twice as great as had previously been thought.

The Earth's surface was not as ideal as had been thought if you considered ever smaller segments; each of these smaller segments revealed new features. The Earth's surface was a patchwork of coupled segments, and the joins between the segments showed. As Méchain commented to a colleague, 'Our observations show that the Earth's curve is nearly circular from Dunkerque to Paris, more elliptical from Paris

---

[4] But improving the precision in spatial localization on the Earth's surface is one of the main reasons for the coming redefinition of the second—see section 12.5.

to Evaux, even more elliptical from Evaux to Carcassonne, then returns to the prior ellipticity from Carcassonne to Barcelona, So why did He who moulded our globe with his hands not take more care …?'; a splendid question, which modern geology can now answer.

As in the scientific advances of the late-19th Century, when scientists attempted to measure something with greater precision, they invariably discovered new science, so with the metric survey. In the late-19th Century, increased precision in measurement led to the creation of quantum mechanics and relativity, in the late-18th Century, increased precision of measurement allowed the creation of 19th Century science.

Further reading

[1] The best place to find further details about science in ancient China is *Science and Civilization in China* by J Needham (Cambridge: Cambridge University Press; 1962); the survey mentioned in this chapter is taken from volume 4 (part 1), pp 42–55. There is a fascinating description of astronomy in Ancient China in volume 3, section 20.

[2] *Le Système Métrique* (The Metric System) by Henri Moreau, published by Chiron (Paris), 1975, is the standard French history of the Metric System and how it evolved from pre-existing systems of units. It is intended for a general readership within schools and colleges as the contents give a French view of these developments in metrology. However, as it was published so long ago, there is no mention of the more recent (last half century) technical developments within this subject.

[3] With regard to the epic surveying adventure of Delambre and Méchain to define the length of the metre, a reader could not do better than consult *The Measure of All Things* by K Adler, published by Abacus, 2004. This is a very readable, non-scientific but detailed account of the precision survey conducted between Dunkerque (Dunkirk) and Barcelona. Even though this volume is long at 370 p of text and an additional 200 p of notes and references, it is informative and easy to follow, with some thought-provoking observations on the difficulties of undertaking detailed scientific work in two countries that are actually at war with each other. It is largely an account taken from a translation of Delambre's autobiography and Méchain's letters

Chapter 4

Measurement in the modern world (II)

4.1 Envy, money, terror, and the Metric System

In French history, 1793 is known as the year of The Terror and not as the year of the metric survey. In the summer of 1793, France was in turmoil. The revolution, which had begun with such high hopes, was at its lowest ebb. Perhaps the surveyor André Méchain had received something of a premonition of what was about to happen in France. On the evening of 10 January 1793, he was in Barcelona and while making detailed astronomical observations he discovered a new comet. This innocent discovery was reported in the local newspaper where the editor felt it necessary to add that a comet was a natural phenomenon whose appearance did not, as the common people believed herald the coming of war, pestilence, or the death of kings. The editor had, however, spoken too soon.

The year continued with the execution of King Louis XVI (21 January), which triggered wars with many of France's neighbours. On 24 February, the National Convention ordered the mobilization of a conscript army of three hundred thousand, but on the following day the poor of Paris looted the food shops. The *sans-culottes* were being egged on in their rioting by the inflammatory journalism of the ultra-radical Jean-Paul Marat (1743–93), who had announced that the exorbitant rises in the prices of goods of first necessity was a plot to destitute the common people, led by a conspiracy of '… capitalists, speculators, monopolists, merchants of luxury goods, the old judges, the ex-nobles'.

Antoine-Laurent de Lavoisier (1743–94) was the greatest chemist of his age. However, he also had a high-profile 'day job', which attracted the envy of the more radical elements of revolutionary politics. Lavoisier was one of the 28 official tax collectors, who collected taxes on behalf of the king on rolling six-year contracts. Needless to say, during their tenure of tax collecting each became wealthy. As was typical of *savants* of that period, Lavoisier used his personal fortune to fund his scientific research. Indeed, he rapidly became an internationally renowned scientist, and was elected to the *Académie des sciences* when only 25 years old. Lavoisier by both his private and his public life excited envy and hatred, and his most implacable enemy was Jean-Paul Marat.

The metric survey was considered a prestige project, and, consequently, the budget for the creation of the Metric System was fixed at 300 000 *livres*, roughly three times the annual operating costs of the entire *Académie des sciences* under the *ancien régime*. Such sums of money angered those sections of society who could not, or would not see the importance of the endeavour. This budget was to be administered by the controller of finance of the *Académie*, the immensely wealthy Antoine de Lavoisier.

The size of this budget raised questions from all sides as to its actual purpose. Many believed that the *savants* were only seeking to enrich themselves. Jean-Paul Marat was even more outspoken, 'those cowardly lackeys of despotism [*will consider the budget, which was public money, as*] a little cake they will share out among confederates'. Although it has to be said that Marat was a dedicated enemy of both the *Académie* and of Lavoisier; the latter having been instrumental in denying him election to the *Académie*, even though he had been an internationally respected doctor before the Revolution.

On 4 May 1793, the Committee of Public Safety fixed the maximum price of wheat hoping to prevent profiteering and to relieve food prices and shortages. Unfortunately, this only precipitated greater hoarding, price inflation, and further food shortages. In addition, to meet the rapidly escalating costs of running the country, the National Convention increased the rate of printing of the new paper currency (*assignats*).

Jean-Baptiste Delambre who was surveying the northern part of the meridian was making good progress, and in the first few months of 1793 had located himself in Paris. Indeed, he had constructed an observation platform above the dome of the Pantheon from which to make detailed observations. From his eyrie, Delambre was able to look down with cool detachment at the food rioters sacking Paris, and at the execution of the king; he was making his measurements above all the strife.

In all the turmoil, the National Convention did not think that the work of the *savants* of the metric commission was proceeding fast enough. To the demand of the politicians that they work more rapidly and make many more approximations in their work, Lavoisier commented 'That would be to substitute a limited and narrow idea for one of the most beautiful and vast conceptions of the human mind.'

Unfortunately for Lavoisier, with the onset of revolutionary violence his enemy Marat found his true calling, becoming a leading populist and rabble-rouser. Marat located himself on the extreme left-wing of the political spectrum in the National Convention; he was as an *ultra* or an extremist, and a supporter of the *sans-culottes* (the poorer and most radical supporters of the Revolution), so called because they usually wore full-length trousers instead of the knee-length *culotte*. Marat was infamous for his inflammatory demagogic journalism, which sent the *sans-culottes* out onto the streets to perform murder and mayhem. But he gained most notoriety by the manner of his death, being assassinated in his bath by Charlotte Corday. However, before his death, Marat had denounced Lavoisier as a wealthy royalist tax collector (which he was) and as someone who sold spoiled tobacco. By his spectacular death (13 July 1793) and the trial and execution of Charlotte Corday which took place a few days later, Marat ensured that Lavoisier would be tried for his life.

Marat may have been silenced, but many still supported his ideas that only the use of the guillotine against traitors and counter-revolutionaries would solve the nation's problems. Maximilien Robespierre entered the Committee of Public Safety in late-July 1793, and four weeks later the general mobilization of France was proclaimed. Yet in all this mayhem, on 1 August the deputies of the National Convention accepted the report of the scientific experts from the Commission on Metric Weights and Measures of the *Académie des sciences* on the creation of the new decimal metric system of weights and measures. The deputies accepted the report, but they said and did nothing, presumably because there were too many more important, and urgent matters pending.

On 4 September, a demonstration of *sans-culottes* demanding higher wages and increased supplies of bread was quickly turned to the advantage of those calling for state-sponsored terror to suppress counter-revolutionaries, and on the following day, they marched on the National Convention. A Jacobin delegate (supporters of the *sans-culottes*) declared 'It is time that equality bore its scythe above all heads. It is time to horrify all the conspirators. So legislators, place Terror on the order of the Day!' The Terror held no hidden mysteries; it was intended to spread paralyzing fear throughout the nation.

Pierre Méchain had gone to Barcelona to work northwards, surveying the meridian. Sadly, before he could complete his measurements in Spain, the Spanish government declared war on France in March 1793. Initially, Méchain was able to use the successful invasion by the Spanish army to proceed with his work into what was now occupied France. Eventually, the French counter-attacked and Méchain and his team were able to smoothly attach themselves to the safety of the French army as their old Spanish protectors retreated southwards. But Méchain's real problem was money; and how to pay his bills, especially after he had been severely injured in a bad fall. Scientists might, with some difficulty, be able to come and go across a war-frontier, but getting money for salaries and living expenses from an inflation-ridden Paris was more of a problem for Méchain. It was Lavoisier who came to the rescue. As Treasurer of the *Académie des sciences*, and one of the richest men in France, he demanded that Spanish banks honour all letters of credit from Paris for the metric survey, 'which concerns the commerce of all nations'.

Lavoisier went further and told Méchain to spare no expense in successfully completing the survey, and that he should not be intimidated by the local difficulties between France and Spain. He managed to establish a chain of communication and finance for Méchain involving letters being passed from the French general commanding the army of the Pyrénées, to his opposite number in the Spanish army; as Lavoisier had commented when pleading Méchain's cause with the military 'The sciences are not at war.'

Sadly, all Lavoisier's formidable skill as a financier and diplomat could not save him from his own political problems in Paris. In November, Lavoisier was imprisoned. The metric commission of the *Académie des sciences* then wrote to the Committee of Public Safety on Lavoisier's behalf, asking that their colleague be released because his work on the detailed measurement of the density of pure water was essential to the successful completion of the project to create the Metric System, and because he had always, in their view, demonstrated zeal in the activities of the commission.

The Committee of Public Safety responded on 23 December by dismissing from the metric commission Lavoisier, Borda, Laplace, Coulomb, and Delambre. Luckily for Laplace, Borda, and Coulomb they were not in Paris when the commission was purged; if they had been in the city, it is unlikely that they would have survived. Delambre was also absent from Paris, surveying in the countryside of the Loire Valley and knew nothing of what was going on back in Paris.

The radical republicans had decided that responsibility, for something as important as constructing a new system of weights and measures could only be entrusted to men known for 'their republican virtues and hatred of kings'. Sadly, Antoine Lavoisier chose to remain in Paris where he lost his head in the spring of 1794, and at whose trial the Procurer General famously retorted to Lavoisier's lawyer, when he asked for a reprieve for his client as he was such an eminent scientist, that 'The Republic needs neither scientists nor chemists; the course of justice cannot be delayed.'

Many years later, Delambre published an autobiography of his work in the metric commission. Here, Delambre blames one particular member of the commission for the political purge. Claude Antoine, *comte* Prieur-Duvernois, known as Prieur de la Côte-d'Or (1763–1832) was a military engineer and a politician during and after the French Revolution and is blamed by Delambre for instigating the *coup de force* of the Committee of Public Safety against certain members of the metric commission.

Apparently, Prieur de la Côte-d'Or was upset at what he considered to be the mocking remarks made against him, during the 'political discussions' of the metric commission, by those members who were subsequently purged by order of the Committee of Public Safety, of which Prieur de la Côte-d'Or was also a member. Given that to be thrown off the commission could have been an automatic death sentence, it was for Lavoisier, as it could be considered a demonstration of a less than perfect zeal for the revolution, we see that neither personal nor national politics mixed comfortably with science during the Terror. Prieur de la Côte-d'Or was a survivor and went on to be the person who presented the completed Metric System to the National Convention one month before it was finally adopted as law on 7 April 1795 [1]; he was only a military engineer but survived the Terror and used the research undertaken by some of the most illustrious mathematicians and *savants* of the 18th Century to further his political career.

The guillotining of Robespierre and his supporters on 28 July 1794 did not bring an end to the Terror; it merely took on a slightly different mask. At the time of Robespierre's death, one *Louis d'or* (a gold coin of the *ancien régime*) had been worth about 75 *livres*, and there were about 6 billion *assignats* in circulation. On 24 December 1794, the National Convention abolished the laws fixing a maximum price for food stuffs and essential commodities. This introduction of liberal market economics resulted in an enormous surge in prices and inflation to such an extent that by the spring of 1795, one *Louis d'or* was worth 4000 *livres* and there were 16 billion *assignats* in circulation. The National Convention was printing huge quantities of paper money, and a state of hyperinflation was being created.

The rapid decline in the state of the national economy was observed by the two surveyors. They were continually writing back to Paris to complain about the rate of price rises, and the total discredit of the revolutionary *assignats*, which they could only use to wrap fish and meat as the local peasants, from whom they sought food knew them to be worthless paper [2].

On 1 April 1795, the *sans-culottes* rioted in Paris over the high price of bread, the availability of bread, and the quality of the bread that was available. The *sans-culottes* attempted to gain entry to the chamber of the National Convention to express their outrage at the consequences of the economic liberalism, but they were dispersed by armed troops. A few days later, on 7 April (18 *germinal* III) the National Convention finally adopted the new Metric System as the only system of weights and measures to be used throughout France. Old, customary units would no longer be used as everyone was expected to use the new, philosophically coherent universal measures.

The Convention now moved quickly to be seen to be doing something about the problem of food shortages and the poor quality of the food that was available, and on 6 May, they ordered the re-distribution of land to peasant farmers and away from the large aristocratic land-owners. Unfortunately, this re-distribution of land was to be undertaken using the units of the new Metric System, which as it was so very new was not at all understood, and its imposition led to a complete breakdown in trust between the people and the politicians.

Between 20 and 23 May 1795, Paris was gripped by food riots. This time, the *sans-culottes* succeeded in invading the National Convention and went so far as to kill a number of deputies and paraded their severed heads stuck on pikes around the Convention, presumably to encourage the republican ardour of the other deputies —'pour encourager les autres'.

The Metric System is an extraordinary achievement, one of the great landmarks in the history of science, but it is something that was born in blood and revolution. The politicians who adopted the Metric System in April 1795 did not adopt it for its philosophical credentials and erudition. The Metric System was adopted because those politicians were in fear for their lives. The adoption of the Metric System was an act of desperation in a world of hyperinflation, and where the prices of basic necessities were being driven ever upwards. The politicians, if they wanted to keep their heads had to be seen to be doing something. The politicians who made the Metric System the only legal system of weights and measures in France, only wanted to preserve their lives. These politicians were not in control of the situation and were desperately trying to manage an unending, horrific crisis. The introduction of the Metric System was perhaps one of the greatest and most glaring examples of the failure of science communication.

4.2 The endgame

The metre gained its legal existence and became a unit of measurement when the National Convention adopted a report from the *Académie des sciences* and decreed the unit of length was to be called the *metre*, and that it was to be 1/10 000 000 of the Earth's quadrant, which it did on 1 August 1793. The two surveyors, Delambre and Méchain, were actually still out in the French countryside measuring the metre, but the politicians had already enacted the legislation giving this new unit a legal existence (the politicians were in a great hurry). While they waited for the meridian to be surveyed and measured, the National Convention issued laws approving a provisional measuring system, and a 'provisional metre' was constructed from the geodetic data already available. A brass standard of the provisional metre was made and it is still preserved in the *Conservatoire des Arts et Métiers* in Paris.

Finally, on 7 April 1795 [18 *germinal*, year (*an*) III in the Revolutionary Calendar, see section 4.3], the National Convention finally and fully adopted the *Académie des sciences*' full recommendation for a decimal Metric System. This new Metric System defined standards for length, mass, area, and volume; and listed the prefixes for multiples and submultiples of units.

This law (*Loi du* 18 *germinal, an III*) adopted the definitions and terms that we still use today. The new measures were officially named *républicaines*. The National Convention decreed that henceforth the new Republican Measures were to be the only legal measures to be used in France. The motto adopted for the new system was Condorcet's famous piece of science communication, *For all time. For all people.*

Despite the enormous difficulties, arising from the political conditions Delambre and Méchain completed their measurement of the distance between Dunkirk and Barcelona in September 1798. Having made the measurements, it was now time to establish the metre and the kilogram. To try and avoid too much nationalist sentiment, it was also felt that this final step would best be done by a committee of *savants* from a number of nations. Unfortunately, France was at war with most of Europe, and the committee which met in Paris contained no representatives from the world's major scientific powers.

Appropriately, the invitations to attend this meeting (the International Commission for Weights and Measures) to fix the new metric standards of length and mass were sent out in the name of the Minister of Foreign Relations, Charles Maurice de Talleyrand-Périgord, who had started the reform of the French system of weights and measures back in March 1790. Now, after having been chased from France in 1792, the great survivor was able to send out on 9 June 1798 a note to diplomats inviting *savants* of countries friendly to France and allied nations to Paris to fix, once and for all the values of the new units.

Representatives came from France, Denmark (a neutral state), the Netherlands (occupied by French troops in 1795), Switzerland (invaded by French troops in 1798 who annexed Geneva), the Kingdom of Sardinia (a French client-state), Tuscany (a dukedom in central Italy belonging to France), the Ligurian Republic (a French client-state centred on Genoa in northern Italy), the Cisalpine Republic (a French client-state centred on Milan in northern Italy), Rome (another French client-state), and Spain (soon to be fully occupied by French troops). The representatives met in Paris to review the survey's findings and to confirm the metric standards. This conference to fix the values of the new metric units is often called the first international scientific conference; however, it was nothing of the sort given the absence of the world's major powers.

Two commissions were then established. The first of these was chaired by the mathematician Jan Hendrick van Swinden (1746–1823) from the Netherlands, who checked and confirmed the results of Delambre and Méchain so that the exact length of the metre could be determined. This commission also ordered the fabrication of platinum and iron standards of the metre and considered the names of the metric prefixes.

The second commission of 1798 was chaired by the professor of mathematics at the University of Berne, Jean-George Trallès (1763–1822), and examined the determination of the mass of the kilogram. They carefully measured the mass of a well-defined cube of distilled water at a well-defined temperature. The size of this container became the standard for the litre and the mass of the water became the standard for the kilogram. Based on the mass of this amount of water, a much smaller and thus more convenient platinum kilogram was then fabricated.

On 22 June 1799, the standards of the metre and kilogram were deposited in the Archives of the Republic. The Metric System was now a fact of life in France and in French-occupied Europe. And it was hoped that the Metric System would, by virtue of its simplicity and technological and, above all its philosophical sophistication spread to other countries to become the universal language of science.

4.3 Avez-vous l'heure s'il vous plait?

Perhaps the most extraordinary change forced upon the French people by the more radical elements of the revolutionary factions, immediately after 1789 was the creation of the French Republican Calendar, or the French Revolutionary Calendar. One can imagine that after throwing off the yoke of the hated monarchy, the people would wish to draw a line under the *ancien régime* and to start a new regime. So, the day the French Republic was proclaimed was decreed to be Day 1 of the new Year 1. However, to go further and to make this new calendar mandatory, as was the case during the Terror was perhaps going too far. Because if you did not use the new calendar or were perhaps just confused by it; your loyalty to the new Republic would be suspect; and in the climate of paranoia, fear and denunciation that then existed, you would soon have lost your head—literally, as opposed to metaphorically.

The French Republican Calendar, which came into existence in late-1793, and lasted until 1805, and which was again used for 18 days by the Paris Commune in 1871, was intended to sweep away the various trappings of the *ancien régime*. The new Republican government sought to institute, among other reforms a new social and legal system, a new system of weights and measures, and a new calendar. Amid nostalgia for the Roman Republic, the theories of the Enlightenment were at their peak and the architects of the new systems looked to Nature for their inspiration.

The new Republican calendar was created by a Commission under the direction of the politician Charles Gilbert Romme (1750–95); it containing a few of the *savants* who already sat on the Commission creating the metric system of weights and measures. The other member of this Commission was the minor poet, actor, and unsuccessful playwright Fabre d'Églantine (1750–94), who was given the task of inventing the names of the new months. The traditional months are named after Roman emperors or pagan gods, and so had to be swept away having no place in the new world of republican logic. The names of the new months were inspired by Nature and were meant to convey something of the character of Nature during that particular period of the year. Charles-Gilbert Romme presented the new calendar to the Jacobin-controlled National Convention on 23 September 1793, which adopted it on 24 October and extended it proleptically to have taken effect on 22 September 1792. Romme, like so many other political radicals, eventually fell foul of the Convention and committed suicide while being taken away to be guillotined. Fabre d'Églantine went to the scaffold on 5 April 1794, apparently distributing auto-graphed copies of his poems to onlookers from the tumbril.

In the Republican Calendar, the year began at the Autumn equinox and had 12 months of 30 days each, which were given new names, principally having to do with the prevailing weather in and around Paris (demonstrating yet again the central-ization of all political power in France). The Republican Year ran as follows[1]:

Autumn
- *Vendémiaire* (from Latin *vindemia* for grape harvest), starting 22, 23, or 24 September
- *Brumaire* (from French *brume* or fog), starting 22, 23, or 24 October
- *Frimaire* (From French *frimas* for frost), starting 21, 22, or 23 November

Winter
- *Nivôse* (from Latin *nivosus* for snowy), starting 21, 22, or 23 December
- *Pluviôse* (from Latin *pluvius* for rainy), starting 20, 21, or 22 January
- *Ventôse* (from Latin *ventosus* for windy), starting 19, 20, or 21 February

Spring
- *Germinal* (from Latin *germen* for germination), starting 20 or 21 March
- *Floréal* (from Latin *flos* for flower), starting 20 or 21 April
- *Prairial* (from French *prairie* for pasture), starting 20 or 21 May

Summer
- *Messidor* (from Latin *messis* for harvest), starting 19 or 20 June
- *Thermidor* (or Fervidor) (from Greek *thermon* for summer heat), starting 19 or 20 July
- *Fructidor* (from Latin *fructus* for fruit), starting 18 or 19 August

Initially, there was a debate as to whether the new calendar should celebrate the Revolution, which famously started in Paris on 14 July 1789 (but had actually begun in Vizille, near Grenoble a year earlier), or the Republic, which was proclaimed in

---

[1] In Britain, people mocked the new Republican Calendar by calling the months: Wheezy, Sneezy and Freezy; Slippy, Drippy and Nippy; Showery, Flowery and Bowery; Wheaty, Heaty and Sweaty.

1792. Eventually, a decree of 2 January 1793 stipulated that the year II of the Republic began on 1 January 1793. The new calendar commemorates the Republic, not the Revolution, that is, it celebrates the politicians, not the people.

In decimal time, the time of day is represented using units which are decimally related. In French Revolutionary Time, the day was divided into 10 decimal hours, each decimal hour into 100 decimal minutes and each decimal minute into 100 decimal seconds. One should remember that the use of this new system was mandatory. Decimal time was introduced into France by a decree of 5 October 1793. The various subdivisions of the 10-hour day were identified in a further decree, 24 November 1793 (that is, 4 *Frimaire* Year *or An* II). The month was divided into three *décades* or 'weeks' of 10 days each, which were simply named: *primidi* (first day), *duodi* (second day), *tridi* (third day), … *décadi* (tenth day). These *décades* were abandoned in April 1802 (that is, *Floréal An* X).

Although clocks and watches were manufactured (see figure 4.1) with faces displaying both standard time as used by humanity with numbers 1–12, and decimal time with numbers 1–10 as used in France (these clocks are highly valued today, because of their exceptional craftsmanship—they contain two mechanisms), it has to be said that decimal time never really caught on. It was not officially used until the beginning of the Republican year III, 22 September 1794, and its mandatory use ceased on 7 April 1795 (18 *Germinal* Year III).

The new calendar was not popular and limped along with the sullen acceptance of the population until some form of stability arose in the government. Napoléon

Figure 4.1. French decimal clock from Revolutionary France. The large dial shows the ten hours of the decimal day in Arabic numerals, with decimal divisions of the hours; while the small dial shows the traditional 24-hour day in Roman numerals. (Image from https://en.wikipedia.org/wiki/Decimal_time#/media/File: Decimal_Clock_face_by_Pierre_Daniel_Destigny_1798–1805.jpg. It has been obtained by the author from the Wikimedia website where it was made available by DeFacto under a CC BY-SA 4.0 licence. It is included on that basis. It is attributed to DeFacto.)

Bonaparte was made Consul for Life in 1799 and in 1804 crowned himself Emperor of the French. And it was Bonaparte who started dismantling many of the more extreme and unpopular innovations of Republican Calendar and the Metric System. The Concordat of 1801 re-established the Roman Catholic Church in France with effect from Easter Sunday, 18 April 1802, and restored the old names of the days of the week with the number and names that existed in the Gregorian Calendar, while keeping the rest of the Republican Calendar, and fixing Sunday as the official day of rest and religious celebration.

The Republican calendar was finally abolished at the end of 1805 by a decree of Emperor Napoléon I with effect from 1 January 1806 (the day after, 10 *Nivôse An* XIV), a little over twelve years after its introduction. However, it was used again during the brief Paris Commune, 6–23 May 1871 (16 *Floréal*–3 *Prairial An* LXXIX), when after the disastrous defeat of the French army by Prussia in 1871, the workers briefly took over Paris and turned it into a workers' commune. The communards showed their extreme left-wing zeal by restoring the trappings and calendar of the Committee for Public Safety. The French made another, final attempt at the decimalization of time in 1897, when the *Commission de décimalisation du temps* was created by the *Bureau des Longitudes* in Paris, with the eminent mathematician Henri Poincaré as secretary. To try and appease some of humanity's incomprehension of the goals of this Commission, they proposed the compromise of retaining the 24-hour day, but dividing each hour into 100 decimal minutes, and each minute into 100 s. The plan did not gain acceptance, and was abandoned in 1900.

During the 20th Century, the idea of decimal time withered away, although in an opening scene of Fritz Lang's 1927 film *Metropolis* the director of the city is shown in his office with a metric clock with ten numbers instead of twelve, and the workers routine is governed by this decimal time. Thereby illustrating the improved efficiency of future industrial society, but also giving a subconscious idea of totalitarianism and fear.

4.4 Falling out of favour with the Metric System

Unfortunately for the world of science and technology but, perhaps, not surprisingly given the instability of French society at that time, no sooner had the decimal Metric System been introduced and made mandatory than the first calls were heard for it to be abolished. Despite the great scientific advance that had been made in creating a rational, coherent system of weights and measures, the ordinary people found the new units, and the new names alien to their culture and experience.

Over a relatively short period after the fall of the Bastille on 14 July 1789, France was ruled by a succession of ephemeral bodies none of which inspired confidence in the people: the National Constituent Assembly 1789–91, the Legislative Assembly 1791–92, the National Convention 1792–95, this period included the Terror of 1793–94, and then from 26 October 1795 (4 *brumaire An* IV) to 9 November 1799 (18 *brumaire an* VIII) France was ruled by the Directory, which was replaced by the Consulate, which lasted until 18 May 1804 (28 *floréal An* XII), and included Napoléon Bonaparte's dictatorship as First-Consul for Life. Then in 1804, First-Consul Napoléon Bonaparte crowned himself Emperor Napoléon I.

This political instability was accompanied by hyperinflation of the currency and civil-war. And the introduction of a new system of weights and measures, which no one understood or trusted, yet would be the basis for land redistribution was not likely to be a success. Not only were the names of the new units introduced in 1795 unconnected to ordinary French culture, but the nomenclature of multiples of the units were derived from classical languages and so were totally unfamiliar to the ordinary people who had to use them. The new system of units would require a profound change to the habits and thoughts of the ordinary people. Put simply, the people for whom the revolution had supposedly been undertaken did not like, nor did they want the new coherent product of French enlightened thinking.

However, it was not only merchants and uneducated peasants who rejected the Metric System. The nation's educated elite clung to their old units just as tenaciously. In 1797, surveyors working for the government had to be reprimanded for not measuring land in metres. In 1799, senior civil servants were still using pre-metric units in all official correspondence; the central Office of Weights and Measures in Paris sent a consignment of new metric standards (rulers) to a provincial office and informed them in the accompanying letter that the package weighed 'soixante livres, poids de marc' (60 pounds of weight). Even the legislators in Paris continued to pass laws in old units; in violation of their own laws.

One year after the metre was made definitive, the first compromise in the, supposedly, inflexible coherence of the Metric System was introduced. On 4 November 1800 (13 *brumaire An* IX), the Metric System was declared to be the sole measurement system for the nation, but the use of the accompanying Latin and Greek nomenclature for multiples and sub-multiples of the units was abolished; the customary names or *mesures usuelles* were re-introduced. After consultation with Delambre and Laplace, the *Académie des sciences* decided that the decimetre would be renamed the *palme* (a hand-width), the centimetre would become the *doigt* (a finger-width) and the millimetre the *trait* (the trace); see table 4.1 for the confusing details.

This initial compromise was instigated by Napoléon Bonaparte himself, who had just escaped from Egypt where his navy had been destroyed by Admiral Nelson at the Battle of the Nile. To honour Bonaparte's return to Paris, the *Académie* struck a medallion in the platinum left over from the manufacture of the first metre standard. The *savants* of the *Académie* fulsomely told the young general that this medallion would last 'almost as long as your glory'.

Then, 13 days after accepting the medallion and its extravagant hopes, Bonaparte seized absolute political power in the *coup d'état* of 18 *brumaire* (9 November 1799), which he would retain for 16 years. One of the first acts of the new First-Consul was to make the greatest of French mathematicians (and the man who had taught and examined him as a young officer), Pierre-Simon Laplace, Minister for the Interior with responsibility for promoting the use of the Metric System. This support for their illustrious colleague prompted the *savants* of the *Académie* to believe that they had been judicious in backing and flattering Bonaparte. However, the academicians soon learned of the compromise over the nomenclature of the metric units, but Laplace

Table 4.1. The changing nomenclature for the units of the Metric System.

Unit of 1 August 1793 The 'early-Metric System'	Unit of 7 April 1795 The 'full-Metric System'	Unit of 4 November 1800 The 'late-Metric System': new quantity, but with old name	Translation	Size (in metres)
Length				
milliaire	myriamètre	lieue	League	10^4
	kilomètre	mille	'a thousand'	10^3
	hectomètre	perche	Pole	100
	decamètre	metre	Metre	10
	metre	palme	Palm	1
	decimètre	doigt	Finger	0.1
	centimètre	trait	Line	0.01
	millimètre			0.001

Unit of 1 August 1793 The 'early-Metric System'	Unit of 7 April 1795 The 'full-Metric System'	Unit of 4 November 1800 The 'late-Metric System': new quantity, but with old name	Translation	Size (in litres)
Liquid volume				
cade	kilolitre	muid	Hogshead	1000
decicade	hectolitre	setier	Barrel	100
centicade	decalitre	boisseau, velte	Gallon	10
pinte, cadil	litre	pinte	Pint	1
	decilitre	verre	Glass	0.1
	centilitre			0.01

Unit of 1 August 1793 The 'early-Metric System'	Unit of 7 April 1795 The 'full-Metric System'	Unit of 4 November 1800 The 'late-Metric System': new quantity, but with old name	Translation	Size (in kilograms)
Mass				
bar		millier		10^3
decibar		quintal	Hundredweight	100

centibar	myriagramme			10
grave	kilogramme	livre	Pound	1
decigrave	hectogramme	once	Ounce	0.1
centigrave	decagramme	gros	Dram	0.01
gravet	gramme	denier	Pennyweight	0.001
decigravet	decigramme	grain	Grain, seed	10^{-4}
centigravet	centigramme			10^{-5}
milligravet	milligramme			10^{-6}

convinced his fellow academicians that this was only a minor problem, and that the purity of the Metric System was in safe hands.

After a few weeks, however, Laplace was dismissed from the Ministry of the Interior, and the post was given by Bonaparte to one of his many brothers, the 24-year old Lucien, who remained Minister long enough to organize the rigged referendum which elected Napoléon as First-Consul for Life; the final step before crowning himself Emperor.

Interestingly, even before the introduction of the Metric System in 1795, the temporary agency established to oversee the introduction of the new units had foreseen the problems of nomenclature. They commented that it would not have been too much of an affront to the new system of units if the old French names for the units had been retained from their very inception, as this would facilitate their acceptance by the people. Evidently, there were a few science communicators who had a sense of what was needed to facilitate such a fundamental change to the nation. A little pragmatism by the authorities, and the new Metric System might have been better accepted.

Although First-Consul Bonaparte had achieved great success in spreading the ideals of the French Revolution throughout Europe by his military conquests, he was not personally convinced of the utility and need for the Metric System. In particular, Bonaparte disliked the inconvenience of surrendering the useful educative quality of traditional measurements in the name of decimalization. The First-Consul also disliked the confusion caused by the French Republican Calendar, which was only used in France.

As Emperor, Napoléon commented in a letter of December 1809 to his Minister of War, Henri-Jacques-Guillaume Clarke, 1st Duke of Feltre, 'I ridicule decimal division'. Similarly, in his memoirs, written during his imprisonment on St Helena, Napoléon wrote, 'The scientists [*who created the Metric System*] had another idea which was totally at odds with the benefits to be derived from the standardization of weights and measures; they adapted to them the decimal system, on the basis of the metre as a unit; they suppressed all complicated numbers. Nothing is more contrary to the organization of the mind, of the memory, and of the imagination .... The new system of weights and measures will be a stumbling block and the source of difficulties for several generations .... It's just tormenting the people with trivia.' In the same source, he went on to mock the global aspirations of the *savants* who had created the Metric System. 'It was not enough for them to make forty million people happy [*the population of France*] ... they wanted to sign up the whole universe.' But that was exactly the idea put forward by Condorcet; the Metric System was to be 'For all time. For all people.' Perhaps the *savants* who had created the metric system had grander plans than Bonaparte himself.

The French now had new units of measurement, but with the old, familiar names; and as one may imagine, their patience was wearing thin. However, the nation was at war so little could be done; and for the moment, France was winning all the battles. Finally, however, by an Imperial Decree of 12 February 1812 (22 *pluviôse An* XX), which was enforced by the government on 28 March 1812 (7 *germinal An* XX), the use of the Metric System was no longer to be mandatory as the old units (those termed *usuelles*) were re-introduced. In effect, the Metric System and the pre-Metric System systems of weights and measures were to be equally valid, and one could pick and choose which system of units one wished to use.

This dual system of weights and measures was the way things stood in France at the fall of Napoléon I in 1815. The restored Bourbon monarchy under King Louis XVIII (the brother of the guillotined Louis XVI) abolished the decimal Metric System on the 21 February 1816 and ordered the exclusive use of the old familiar, or *usuelles* system of units for all commerce. The Metric System was seen by Louis XVIII as a product of the French Revolution and its bloody aftermath, which had organized the death of his elder brother and subverted the natural monarchical order of France. The restored monarchy was not greatly interested in the philosophical coherence of the Metric System; as far as they were concerned, it was merely a product of radical politics. This was not an unreasonable conclusion given the political conditions of 1789–95.

However, the French Revolution did more for modern science than simply introducing order and rigour into weights and measures in the form of a 'pseudo-political' Metric System. This period also led to the creation of a new, more technological world of politics, political science, and international relationships, marking the beginning of a true, elitist scientific community. This new scientific community, which gave rise to the new profession of 'scientist' (*scientifique*) abolished the concepts of the natural philosopher and *savant* that had been built upon the common ideals of solidarity, which had existed in the community of natural philosophers, and in the spirit and culture which came from the Enlightenment. These reforms transferred all authority for matters related to weights and measures to the new elite scientific community.

In matters of science, the revolutionary legislative activity of the 1790s broke completely with the past. The French Revolution imposed a logic of numeration and quantization on the whole nation, a new tyranny of numbers. This obsession with numbers, accurate time-keeping and statistics is still apparent in the private and public lives of everyone today.

It is perhaps not surprising that after the turmoil of the preceding years, the changes made to the Metric System by First-Consul Bonaparte caused the nation's system of weights and measures to rapidly descend back into the anarchy that had existed when Talleyrand had first proposed a reform of the system back in March 1790. This confusion can be clearly seen in table 4.1, which shows how the nomenclature for some of the fundamental quantities, used by the whole nation, had changed over a seven-year period.

The beautifully conceived and philosophically coherent Metric System had been introduced to the country at the wrong moment, and in too hurried a fashion. No one had attempted to explain to the people what it was all about, and the people did not trust the *savants* or the politicians so they naturally took against the Metric System.

The French Revolution changed many institutions and forced man to re-assess many of his ideas. For metrology, gone was the idea that 'man was the measure of all things', the figure and dimensions of man disappeared from the heart of the new

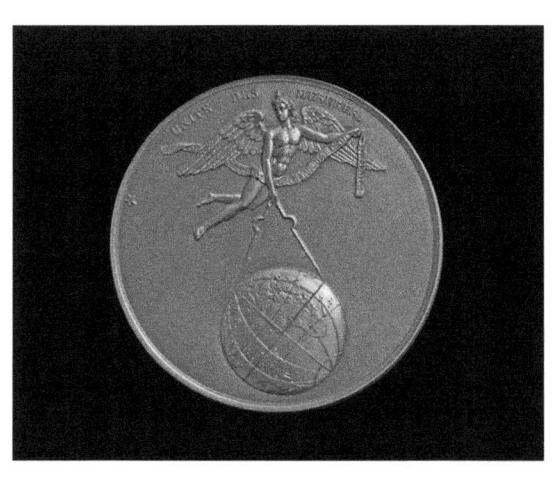

Figure 4.2. A medallion struck in Paris in 1840 to mark the final introduction of the Metric System into France and to act as a souvenir to posterity of the manner in which the metre was determined in 1799; the allegorical figure is measuring one quadrant of our planet. The reverse side of this medallion bears Condorcet's famous rallying call for the Metric System, *A tous les temps. A tous les peuples.* (This image is reproduced with the permission of the BIPM, which retains full international copyright.)

system of weights and measures. Protagoras had told us that the dimensions and proportions of the human body are mirrored in the dimensions and proportions of Nature. All this was swept away by the Cartesian ratiocination of the French Revolution.

The next change to French weights and measures came under the last monarch that France has known. On 4 July 1837, King Louis-Philippe ordered the abolition of the Imperial Decree of 1812 and forbade, from the beginning of 1840 the use on any units of measurement other that those associated with the decimal Metric System of 1795 (figure 4.2, which shows a medallion struck to commemorate the final introduction of the Metric System into France). France had finally adopted the Metric System. However, even at this late-date, there was still political opposition to the Metric System, and the Chamber of Deputies and the second-house of the parliament, the Chamber of Peers, forcibly pushed through the legislation needed for France to finally adopt the Metric System. Not surprisingly, there was still opposition to the Metric System among the ordinary people of France; however, apart from one act of rebellion in the countryside between Paris and Lyon, at Clamecy where the *Préfet* had to call out the militia to break up a band of farmers who had refused to accept the new metric units, future opposition to the new units was consigned to satirical newspapers.

Further reading

[1] *Le Système Métrique* (The Metric System) by H Moreau, published by Chiron (Paris), 1975, is the standard French history of the Metric System and how it evolved from pre-existing systems of units. It is intended for a general readership within schools and colleges as the contents give a French view of these developments in metrology. However, as it was published

so long ago, there is no mention of the more recent (last half century) technical developments within this subject.

[2] With regard to the epic surveying adventure of Delambre and Méchain to define the length of the metre, a reader could not do better than consult *The Measure of all Things* by K Adler, published by Abacus, 2004. This is a very readable, non-scientific but detailed account of the precision survey conducted between Dunkerque (Dunkirk) and Barcelona. Even though this volume is long at 370 p of text and an additional 200 p of notes and references, it is informative and easy to follow, with some thought-provoking observations on the difficulties of undertaking detailed scientific work in two countries that are actually at war with each other. It is largely an account taken from a translation of Delambre's autobiography and Méchain's letters.

Chapter 5

Creating the language that is science

5.1 Dividing apples with oranges to make … something different

Earlier, we encountered the idea of John Wilkins for a universal measure that would serve to create an entire system of weights and measures. In fact, Wilkins was seeking to create a universal philosophical language, which could be understood by all men.

By the end of the 17th Century, such ideas of a rational universal language were very much part of the European *zeitgeist*. In 1666, the German polymath Gottfried Wilhelm Leibnitz (1646–1716) published his *Dissertatio de arte combinatoria* in which he claimed that a true philosophical language would be able to analyse all possible concepts into their simplest elements; into what Leibnitz termed 'the alphabet of thought'. In such a philosophical or, as we would say today, a scientific language, a proper symbol should indicate the nature of the animal, phenomenon or whatever it was naming. In other words, it was a language, which could define that thing or that phenomenon by means of that thing's appearance or that phenomenon's intrinsic properties. Such an idea is premised upon the concept that all ideas can be deconstructed into primitive notions or base units, and that relations exist between these base units.

Leibnitz was writing in a century which had attempted the construction of many universal philosophical languages and so presupposed that a complete enumeration of human knowledge could be achieved. The question that arises immediately is how a relatively small number of base units could be manipulated or combined to produce a true universal scientific language capable of describing all of Nature?

The philosophical languages of the 17th Century were attempting to reform the natural (vernacular) languages by simplifying the complex multiple meanings of some words and concepts. Consider an attempt at learning the definitions of all the words in a dictionary, or of attempting to comprehend all of science. In the dictionary you will find every word defined in terms of other words. In your determination to learn the meaning (or meanings) of every word, you may find that you need to consult the definitions of the words employed in the definitions of other words. Indeed, you soon realize that your initial attempt at learning the meaning of each word in the dictionary is futile. In fact, it is a circular task, because the dictionary contains only a closed set of words, finite in number, that enable descriptions of the meanings of each other. If you do not already have in your mind a set of basic words whose meanings you know independently, without the need of words to define them, you will remain forever in a continuous loop with your dictionary. For the same reason, the philosophical languages did not reform and simplify English, but they did give us the thesaurus.

At the time of the French Revolution, *savants* who were familiar with the ideas of Wilkins assumed that the new fundamental unit of length, the metre, could be used to define all the scientific and technological concepts required by their society. This metre was to be defined from the dimensions of the Earth, as one ten millionth of a quadrant of the Earth's circumference. This universal unit of measurement, the metre, is in fact one of the seven base units of the International System of Units (the SI).

Having defined the basic unit of length, l, distances could now be expressed in this unit. Then, to define an area of land, a two-dimensional quantity, you simply multiplied two distances, each expressed in metres. Similarly, three spatial dimensions define a volume as so many cubic metres.

These examples show how the single base unit of length can be used to construct a number of other essential components of a system of weights and measures. To go further, we may, for example, assume that the density (the mass of a known volume of a substance) of pure water is well-defined as one gram for each cubic centimetre; one can then define a base unit of mass as the weight of a precisely known volume of pure water. The kilogram was originally defined as the mass of 1000 cm^3 (a cubic box where the sides are each of 10 cm) of pure water.

But what happens when we wish to consider the combination of length with other quantities which are essential in even some of the simplest concepts of technology; for example, how does one introduce time into a system of mechanical quantities?

The speed, or velocity of a planet flying through space, or of a man walking is defined in terms of distance and time, yet how do we combine these two different base units? One might think of these two quantities as being as different as apples and oranges, so how can they be divided or multiplied together; they certainly cannot be added or subtracted? It is the mathematical definition of a unit that allows us to manipulate distance and time and generate new ideas such as the concept of speed, and of acceleration which is speed or velocity per unit time.

First, consider what we mean by a unit. Any value of a physical quantity, Q, may be expressed as the multiplied product of a unit $[Q]$ and a purely numerical factor or value (a simple number). Written algebraically, we have $Q =$ (a number) $\times [Q]$, where $[Q]$ is the unit; for example, there are a certain number of metres or seconds, $Q_{length} = 10$ m or $Q_{time} = 10$ s. Written in this manner, the multiplication sign between the number of metres and the unit of distance (the metre) or time (the second) is omitted for simplicity.

This convention of expressing a quantity (any quantity) as a unit and a numerical factor is used throughout science and is referred to as quantity calculus. When units are being manipulated, one may only add like terms, as with apples and oranges, but all units must be manipulated algebraically. When a unit is divided by itself (that is, metres/metres or seconds/seconds), the division yields a dimensionless number, which is one (1) and so intrinsically without dimension, or a unit. When two different units are multiplied or divided, the result is always a new unit, referred to by the combination of the individual units. For instance, in the SI, the unit of speed or velocity is metres per second; that is, metres/seconds or m/s or m s^{-1}. This new unit is neither length nor is it time, but length divided by time. Only the numerical values are actually divided; they are simple numbers and so can be divided; the two original units are distinct, and cannot be divided, but are left as a new unit; metres divided by seconds.

Length and time are base units; that is, they cannot be decomposed into simpler components, but the new unit of velocity is said to be a derived unit and may be deconstructed into base units. In the same way, the density of water we encountered above in the definition of the base unit of mass is also a derived unit. Density is defined as the mass of a known volume or mass per unit volume and is composed of two base units, the base unit of mass (kilogram) and the base unit of length (metre), which as we are dealing with a volume is cubed (kilogram/metre^3 or kg m^{-3}).

As science and technology advanced in the 19th Century, the new profession of scientists understood how the various manifestations of, for example, heat and work were all related to the concept of energy, and how this idea related to the established base units of length, mass, and time. In fact, today we have seven base units which may be combined to explain every known scientific phenomenon, and which would be used to comprehend scientific discoveries that have yet to be made. That is, it is through these seven base units that the true universal language, the language of science is formulated.

By convention, all physical quantities are organized into a system of dimensions. Each of the seven base quantities used in the modern SI is regarded as having its own dimension, which is symbolically represented by convention as a single *sans serif* roman capital letter (this is some of the dogma of the SI). The symbols used for the base quantities or pre-May 2019 base units[1], and those which are used to denote their dimensions are given in table 5.1.

---

[1] Prior to May 2019, one would have said that the SI was based on seven base units, from which all derived units are constructed (as explained in this book). However, the recent changes to the SI, that is, the creation of the Quantum-SI, which is the main topic of the later chapters of this book has complicated the definitions of the seven base quantities that make up the Quantum-SI. In the recent redefinitions, four of the seven SI base units—the kilogram, ampere, kelvin, and mole—were redefined by setting exact numerical values for the Planck constant (h), the elementary electron charge (e), the Boltzmann constant (k_B), and the Avogadro constant (N_A). The second, the metre, and the candela were already defined by physical constants and were subject to a refinement of their definitions. These new definitions have created a Quantum-SI, which improves the SI without changing the value of any units, ensuring continuity with existing measurements. These changes were brought about by the international community, as explained in chapter 10, and represent the culmination of decades of research. The pre-May 2019 (old) base units are now defined by the fixed values of fundamental constants of Nature, rather than by a unit with an attached, specified experimental set of conditions. The only base unit with a definition close to the old specified experiments, is now the definition of the candela; although this quantity is somewhat redundant as it could be defined in terms of the new definition of thermodynamic temperature.

Table 5.1. Base quantities and their dimensions, and the base units of the SI.

Base quantity	Symbol of base quantity	Dimensional symbol	SI base unit	Symbol
Length	l	L	metre	m
Mass	m	M	kilogram	kg
Time	t	T	second	s
Electric current	i	I	ampere	A
Temperature	T	Θ	kelvin	K
Amount of substance	n	N	mole	mol
Light intensity	I	J	candela	cd

All other quantities, that is, all the phenomena known to modern science, are derived from these seven base quantities using the well-established equations or laws of physics, and are called derived quantities. As outlined above, the dimensions of the derived quantities are written as products of powers of the dimensions of the base quantities using the equations that relate the derived quantities to the base quantities[2].

Dimensional analysis is a powerful tool in understanding the properties of physical quantities independent of the units used to measure them and is the manner in which we use the language of science to define and measure all nature. Every physical quantity is some combination of the quantities in table 5.1. Dimensional symbols and exponents are manipulated using the rules of algebra; for example, the dimension of area is written as L^{-2}, the dimension of velocity as $L\,T^{-1}$, the dimension of acceleration (the rate of change of velocity with respect to time) is written as $L\,T^{-2}$ (metre per second squared), and the dimension of density as $M\,L^{-3}$ (kilogram per metre cubed).

Dimensional analysis is routinely used to check the plausibility of newly derived equations, the design of experiments, and the results of calculations in engineering and science before money and effort is expended on detailed measurements. In this way, reasonable hypotheses about complex physical situations are examined theoretically, to see if they merit subsequent testing by experiment. And, it is also the means by which one seeks to determine appropriate equivalent values for a quantity in another system of units. For example, how you convert from the value of a quantity in metric units to the equivalent quantity in British customary units; metres/second to miles/hour or joules (the SI derived unit of energy, symbol J, where J is equivalent to kg m^2/s^2; that is, $L\,M^2\,T^{-2}$) to British Thermal Units or BTU. [A customary unit of energy equal to about 1055 joules. A BTU is approximately the

---

[2] In general, the dimension of any quantity Q in science is written in the form of a dimensional product

$$\text{Dimensions of } Q = L^{\alpha}M^{\beta}T^{\gamma}I^{\delta}\Theta^{\varepsilon}N^{\zeta}J^{\eta}$$

where the exponents α, β, γ, δ, ε, ζ, and η are generally small whole numbers (integers), they can be positive or negative, or even zero, and are termed dimensional exponents. This simple formula defines the make of all things.

amount of energy needed to heat one pound (0.454 kg) of water, which is exactly one tenth of a UK gallon or about 0.1198 US gallons, from 39 °F to 40 °F or 3.8 °C to 4.4 °C]. Thus, dimensional analysis is the means of translating between the various dialects of the single universal language of science.

Consider the concept of force, that is, something experienced by an object to make it change its speed or velocity through, for example, acceleration. In the SI, the unit of force is the newton (symbol, N), named after Isaac Newton in recognition of his fundamental work in mechanics. The newton is equal to the force required to accelerate a mass of 1 kg at a rate of 1 m s^{-2}. In dimensional analysis, we use Newton's famous formula (his Second Law of Motion) where force (F) is given as being equal to a mass (m) multiplied by acceleration (a), that is $F = ma$, where we are multiplying m (kilogram) by an acceleration a (metre/second^2), and the dimension of the newton is found to be M L/T^{-2} or M L T^{-2} or kg m s^{-2}. The newton is derived only from the base units of mass, length, and time, and so could have been derived by the *savants* of the late-18th Century.

Dimensional analysis was known to Isaac Newton who referred to it as the *Great Principle of Similitude*. The 19th Century French mathematician and Egyptologist Joseph Fourier (1768–1830) made important contributions to dimensional analysis based on the idea that physical laws like Newton's law, $F = ma$, should be independent of the systems of units employed to measure the physical variables. That is, the laws of Nature and fundamental equations should be equally valid in the Metric System of units and in a non-metric system of units. When converting between these two systems of units, we need only be aware of the mathematical factors needed to convert between the base units to convert the entire quantity from one system to another. Thus, one should take care never to mix systems of units, as the consequences could be disastrous (see below). But there is nothing stopping one defining force in ancient Egyptian units; distance would be in terms of the royal cubit (about 0.525 m), mass would be in *deben* (about 0.015 kg), and time would have been in *unut* (the hour, which is identical with our modern hour). Fourier's ideas show how each of these base units would need to be converted to SI base units to convert the ancient Egyptian unit of force (a *pharaoh*?) to the newton.

5.2 The consequences of mixing units

Perhaps one of the costliest examples of not following Fourier's ideas on dimensional analysis and of inadvertently mixing units occurred on Mars on 23 September 1999, when the Mars Climate Orbiter satellite was lost during a manoeuvre to place it in an orbit around the Red Planet.

After the long-crossing of interplanetary space, the satellite's controllers would have needed to significantly slow down the satellite for it to safely enter the Martian atmosphere. It is believed that it was this braking process, which led to the satellite's loss. To slow down a body in motion requires the application of a force of the same order of magnitude as the force generated by the body's forward motion, but applied in the opposite direction.

Apparently, the force needed to slow the satellite down for it to enter a stable orbit around Mars was calculated in one set of units, but when the command was sent to the satellite to ignite the braking thrusters, it was applied in a different set of units. The two sets of software, on Earth and on the satellite hurtling towards Mars, did not realize that they were trying to communicate in different scientific units; and instead of entering a stable orbit well above the surface of the planet, it attempted to enter an orbit much closer to the surface and disappeared.

The principal cause of the disaster was traced to a thruster calibration table, in which British customary units instead of SI units had been used to measure force. The navigation software expected the thruster impulse data to be expressed in newton seconds (the SI unit), but the orbiter provided the values in pound-force seconds. This confusion in units caused the electric impulse to be interpreted as roughly one-fourth its actual value.

In the SI system, the newton is the unit of force and is equal to the force required to accelerate a mass of 1 kg at a rate of 1 m s^{-2}. The pound-force (lb_f from pound with a symbol for force) is a unit of force in the system of units loosely term British Imperial, or customary units. The pound-force is equal to the force exerted on a mass of one *avoirdupois* pound on the surface of Earth.

The acceleration of the standard gravitational field (g) and the international *avoirdupois* pound (lb_m) define the pound-force as $1 \text{ lb}_f = 1 \text{ lb}_m \times g = 1 \text{ lb}_m \times 32.174 \text{ ft s}^{-2}$, which on converting to the SI is equal to $0.454 \text{ kg} \times 9.806\,65 \text{ m s}^{-2} = 4.448$ newtons. (This is the factor which caused the loss of the satellite.)

The disappearance of the satellite illustrates an important point; even after two hundred years of the decimal Metric System, the world of science and technology is full of different systems of units, and converting between them requires attention.

5.3 Derived units

The base quantities of the SI given in table 5.1 are combined to generate derived units, which are products of powers of base units without any numerical factors (other than 1). Such a set of coherent derived units is, in principle, without limit and they represent the means by which all the phenomena of nature are described. Table 5.2 lists some examples of derived quantities and the corresponding coherent derived units expressed in terms of the base units of the SI.

In this way, the base units of the SI are combined to produce a language capable of describing Nature, the make of all things.

Some important derived units are given a specific name, usually to honour the scientist most closely associated with that quantity. Some of these named derived units are given in table 5.3.

Thus, by using the laws of physics as a grammar, and the base units as expressions or words, we may construct a language that allows us to make predictions about phenomena that have not yet been identified, but should be observable. Looking at these tables, a scientist could ask questions such as: what happens if a force tries to twist a body instead of pushing (repelling) or pulling (attracting) it? or what happens if a force acts upon an area, not simply along a line?

Table 5.2. Derived quantities.

Derived quantity	Derived unit	Representation
Area	square metre	m^2
Volume	cubic metre	m^3
Velocity or speed	metre per second	m/s or $m\ s^{-1}$
Acceleration	metre per second squared	m/s^2 or $m\ s^{-2}$
Density	kilogram per cubic metre	kg/m^3 or $kg\ m^{-3}$
Surface density	kilogram per square metre	kg/m^2 or $kg\ m^{-2}$
Specific volume	cubic metre per kilogram	m^3/kg or m^3kg^{-1}

Table 5.3. Named derived quantities.

Derived quantity	Derived unit (symbol)	Representation
Frequency	hertz (Hz)	s^{-1}
Force	newton (N)	$m\ kg\ s^{-2}$
Pressure	pascal (Pa)	$m^{-1}\ kg\ s^{-2}$
Energy or work	joule (J)	$m^2\ kg\ s^{-2}$
Power or light intensity	watt (W)	$m^2\ kg\ s^{-3}$

In the first case, one defines torque, which is a force that tries to rotate a body. In the second case, a force acting over an area would be newtons per square metre, and from the tables we see that this quantity would be $m\ kg\ s^{-2}/m^2$ or $m^{-1}\ kg\ s^{-2}$, which is the definition of pressure. Pressure is nothing more than the force exerted by something (a gas or a fluid) upon a well-defined area. The British customary unit for pressure is pounds per square inch, which gives a clear indication of pressure as a force upon an area.

As for the reasonableness of phenomena that have not yet been observed; one first has to consider the magnitude of the units and a dimensional analysis to see if such a new phenomenon would be observable. A relatively recent example would be the pressure exerted by light, radiation pressure. Could it exist? Is it measureable? The answer is in the affirmative, and it was discovered that the radiation of the Sun exerts a pressure of less than a billionth of an atmosphere on the Earth's surface. But it was an examination of the existing language of science, which suggested to, and allowed individuals to look for this relatively new phenomenon.

5.4 A final comment on the value of a quantity

The biblical text known as the *Book of Revelations* can tell us something important about the nature of units of measurements.

The concept of a sacred architecture comes from Saint John's vision of the Heavenly Jerusalem given in verse 17 of chapter 21 of the *Book of Revelations*. The King James (Authorized) Bible tells us about the dimensions of the coming

New Jerusalem, 'And he measured the wall thereof—an hundred and forty and four cubits, according to the measure of man, that is, of the angel.'

Remember that the cubit was a unit of length common to the ancient Mediterranean world, equal to the length of a man's forearm. We see immediately from this text that the author is in fact referring to the 5th Century BCE, pre-Socratic Greek philosopher Protagoras' well-known comment that 'man is the measure of all things', but we are also given the actual height of the walls of the heavenly city, 144 cubits. This dimension has over the last two millennia, inspired many individuals; particularly, architects.

The great gothic Cathedral of Amiens in northern France, where construction began in 1220, is built to a height of 144 *pieds Romans* (that is, 42.3 m or 138.8 British feet). The nearby Cathedral of Beauvais on the other hand, where construction began in 1225, is built to a height of 144 *pieds du Roi* (that is, 48.5 m or 159 British feet). These two mediaeval units, Roman feet (*pieds Romans*) and royal feet (*pieds du Roi*), are different. The fact that royal feet are longer than Roman feet means that the Cathedral of Beauvais was higher than the Cathedral of Amiens, which may well explain why it was an unstable building that partially collapsed in 1284, while the Cathedral of Amiens has never fallen down.

As far as the mediaeval architects of these two neighbouring cathedrals were concerned, it was the numerical value of the height of the City of God that was important. One hundred and forty-four units was to be the height of the cathedrals because that was the height of the City of God given in Revelations **21:17**; it appears not to have mattered much which units were actually adopted. (Remember: any value of a physical quantity, Q, may be expressed as the product of a unit [Q] and a purely numerical factor or number.) The gothic architects of northern France were only interested in the numerical factor of this physical quantity taken from the *Book of Revelations*; as far as they were concerned, the unit was irrelevant.

This addiction to the value of a unit is numerology. That is, attempting to find some secret hidden in a particular number. What precisely is the meaning of 144? When you put the heights of the two cathedrals into British feet or metres, any mystical significance disappears in the very different light of the British Industrial Revolution and the French Enlightenment.

Conversely, the ancient Sumerians had little concept of pure numbers. Our earliest recorded lists of objects come from Ancient Sumeria, and when we look at these ancient lists on clay tablets (about five millennia old) we see that the scribes used the same metrological symbol or unit as many times as was required by the value of the numerical factor before the unit. Thus, instead of writing six oxen as the Sumerian form of the number six followed by a schematic of an ox, the scribes simply drew the schematic of the ox six times.

The ancient Sumerians used only the unit part of the definition of a quantity, while the mediaeval craftsmen only used the numerical part of the definition of a quantity. Neither is the correct approach. One can readily imagine the evolution of the Sumerian usage to a more sensible, modern approach arising because the lists were getting to be very long and tedious to compose, and those clay tablets were so small. The magico-Christian architects, on the other hand, got caught up in looking for mystical significance in the numbers mentioned in religious-poetic texts of late-Antiquity, and as a consequence they lost their way in numerology and their cathedral collapsed.

Further reading

[1] A fascinating, and very readable account of the philosophical languages proposed by John Wilkins and Gottfried Leibniz, and from which modern ideas of rational systems of weights and measures evolved is *The Search for the Perfect Language* (1997), by Umberto Eco; Fontana Press, London.

Chapter 6

What was not in the original Metric System?

The Metric System of April 1795 introduced standards for those quantities, which were most familiar to the *savants* of the time. Length was Wilkins' original 'universal measure', and from length one could derive area and volume. Time was well established by the late-18th Century, where the most accurate clocks had allowed the Earth to be measured and navigated; creating a map of lines of latitude and longitude.

The Metric System was introduced at the start of the Industrial Revolution. Steam engines, for example, were becoming well-known devices, and *savants* had been experimenting with electricity. So, it would be reasonable to inquire as to the status of the quantities of heat and of electricity at the time of the original Metric System.

6.1 Energy, work, and power

The great problem for the *savants* of the 18th Century was that there was no technical language of science with which they could study Nature. Indeed, it was these individuals who began to create such a language. Today when we talk of energy, we know that it comes in many forms: light, heat, mechanical energy, and electrical energy. But in 1795, *savants* could only suppose that there might be an equivalence between the rise in temperature noted when two mill wheels ground corn, and the rise in temperature noted when one stood in sunlight.

As far back as 600 BCE, the Greek philosopher Thales of Miletus had suggested that everything in the world is, in some manner, conserved. As it happens, these philosophers were not considering the conservation of energy, as we term it today, but the conservation of matter, which was something they could observe and attempt to quantify. It was Galileo, who in 1638 published an analysis of his observations on the motion of pendulums in which we began to see the modern concept of energy conservation, or of energy changing from one form into another.

The earliest ideas on the conservation of energy were put on a mathematical basis by Gottfried Leibniz who formulated the concept of an 'animation' associated with the motion of an object (what today we call kinetic energy, which is proportional to the mass of a moving object multiplied by its velocity squared) and who first proposed that this 'animation' was conserved in a system of colliding masses. That is, a moving object could transfer energy to another object during a collision. Leibniz termed this conservation of energy, in a strongly interacting mechanical system the *vis viva* or living force.

During the 18th Century, it was suspected that the heat generated by mechanical friction, was another form of *vis viva*. In 1783, Antoine Lavoisier and Pierre-Simon Laplace investigated the competing theories for energy flow between solid objects; that is, the *vis viva* and the idea that energy flowed like a fluid between two interacting objects, the caloric theory, which had been formulated by Lavoisier but who did not live to see his theory overthrown. However, it was the measurement of the large amounts of heat generated during the boring of cannons that finally convinced the *savants* of the day that mechanical motion could be converted into heat, and that friction was a route for the loss of that energy which had not been considered by Leibniz in the formulation of his *vis viva*. These studies of the heat generated by the mechanical action of drilling cannon were undertaken by Sir Benjamin Thompson (1753–1814) an American-born, British scientist and inventor, whose challenges to the then established physical theory of heat were a central contribution to the 19th Century revolution in the physics of heat, and the transfer of heat.

The demonstration of the mechanical equivalence of heat by Thompson was a key stage in the development of the modern principle of the conservation of energy. The caloric theory maintained that heat could neither be created nor destroyed, but conservation of energy entails the more subtle principle that heat and mechanical work are actually interchangeable. In 1807, Leibniz's *vis viva* started to be known as 'energy' in consequence of the ideas of the English polymath Thomas Young (1773–1829), who also found the time to decipher the Egyptian hieroglyphs on the Rosetta Stone.

Today, we formulate these ideas of energy conservation as the First Law of Thermodynamics, which may be stated in a number of ways, e.g. 'Energy can neither be created nor destroyed. It can only change forms.' Although the First Law laid the basis for our understanding of heat or energy transfer, it was in France early in the 19th Century that one of the first serious theoretical investigations of the new inventions of the Industrial Revolution was undertaken, and which led to a fuller understanding of the nature of heat and energy. This fuller mathematical statement is called the Second Law of Thermodynamics.

Nicolas Léonard Sadi Carnot (1796–1832) was a French military engineer who made the first theoretical model of how steam engines function. Although everyone could clearly see that in a steam engine you put energy, by burning coal or wood, into the engine to heat the water to generate the steam, which was then used to perform work, no-one knew how the quantity of work generated by the engine related to the amount of energy put into the engine. Consequently, it was not possible to answer *a priori* questions such as: was there a means of making the steam engine more efficient? or could one use fluids other than water in such a device? Without a mathematical model of the steam engine, to answer these questions would have required the construction of many different types of steam engine, which would have been expensive and time-consuming.

In 1824, Carnot published his *Reflections on the Motive Power of Fire*, which gave the first coherent theoretical account of how steam engines function (this general class of devices are called heat engines). Carnot showed how the energy put into the heat engine is related to the mechanical energy generated by that heat engine, and which is available to do work. Today, we call this analysis the Carnot Cycle and it laid the foundations for the Second Law of Thermodynamics.

By idealizing the steam engine, Carnot arrived at clear answers to the above questions. He showed that the efficiency of an idealized steam engine equals the difference in absolute temperature between the hot reservoir (the temperature of the steam) and of the cold reservoir (the temperature of the water in the boiler) divided by the absolute temperature of the hot reservoir. Carnot's analysis revealed that the maximum efficiency of a steam engine was only about 38%. This efficiency of converting the energy released by burning coal to heat water, and then using the steam generated to perform mechanical work, may seem modest but it did power the Industrial Revolution.

Carnot's analysis of how the energy put into the steam engine is converted into work was the first to demonstrate that perpetual motion machines are not possible. His work laid the foundations for the concept of entropy, which determines that thermal energy always flows spontaneously from regions of higher temperature to regions of lower temperature, and it also tells us why systems containing energy run down, and that some processes are irreversible.

Sadi Carnot also laid the foundations for the theoretical formulation of complex and abstract scientific problems, which is a constant of modern science and engineering. Unfortunately, Carnot's book received very little attention from his contemporaries; he died young of cholera, and was quickly forgotten. Carnot's work only began to have a real impact when modernized by Émile Clapeyron in 1834 and then further elaborated upon by Rudolf Clausius and Lord Kelvin, who together in 1865 derived from it the full nature of entropy and the modern formulation of the Second Law of Thermodynamics.

Energy was not included in the original Metric System, but how might we consider energy within the framework of the original Metric System?

Energy is the quantity that allows a system to do work. And work is defined as an amount of energy, which may be transferred from one system to another system by a force acting over a distance. Thus, energy and work are interchangeable. In the scientific form of the Metric System, the derived unit of energy and work is the joule (symbol J) named after the British scientist James Prescott Joule (1818–89), see figure 6.1, and is defined as newton metres, that is, a force, defined in units of newtons acting over a distance of one metre (i.e. N m). This definition of energy comes from the definition in Newton's Second Law of a force $F = ma$, as $M\,L\,T^{-2}$, which when it acts over a distance L gives us energy as $M\,L^2\,T^{-2}$ or $kg\,m^2\,s^{-2}$. We can

Figure 6.1. A photograph of James Prescott Joule, the Salford brewer, and amateur scientist who defined and quantified energy (image from: https://en.wikipedia.org/wiki/James_Prescott_Joule#/media/File:Joule_James_sitting.jpg; it has been obtained by the author from the Wikimedia website where it is stated to have been released into the public domain. It is included on that basis.)

see the universal nature of this definition of energy when we consider Einstein's famous definition of the energy derived from nuclear processes, where the energy released is equal to the mass lost multiplied by the velocity of light squared; that is $E = mc^2$ or M $(L/T)^2$ which is M L^2 T^{-2} . Power is the rate at which work is done; that is, energy divided by time, or joules/seconds, or M L^2 T^{-2}/T which is M L^2 T^{-3}.

Such simple relationships involving mass, distance, and time would have been familiar to the *savants* of the late-18th Century, as they involve the fundamental base units of the Metric System of 1795. Yet, these same *savants* would not have known of these relationships because the physics underlying them was not fully formulated until the middle of the 19th Century, which is why we say that the base units of the universal language of science should be able to explain all scientific observations— those phenomena of which we are aware, and also of those phenomena of which we have yet to become aware.

6.2 Electricity

In the same way that the study of heat and energy had not reached a sufficiently mature stage to allow the *savants* who formulated the Metric System to propose a base unit for temperature in 1795, the study of electricity was at an even more rudimentary stage. Indeed, the study of electricity, electrical phenomena, and magnetism in the 1790s had more in common with parlour tricks than laboratory investigations.

The two names associated with the earliest investigations of the nature of electricity are the pious, conservative, Italian medical doctor Luigi Alyisio Galvani (1737–98), and another Italian, the *savant* Alessandro Giuseppe Antonio Anastasio Volta (1745–1827).

In 1791, Galvani famously discovered that the leg muscles of dead frogs twitched when they came in contact with an electrical spark. According to popular versions of the story, Galvani was dissecting a frog at a table where he had previously been investigating discharges of static electricity. Galvani's assistant touched an exposed sciatic nerve of a dead frog with a metal scalpel, which had apparently picked up a residual static charge. At that moment, the two men saw the leg of the dead frog kick as if it were alive. Such laboratory-based observations made Galvani the first to consider the possible relationship between electricity and animation; that is, the creation of life itself and the possibility of the re-animation of dead tissue.

Galvani used the term 'animal electricity' to describe the force that animated the muscles of the dead frog. Along with many of his contemporaries, he regarded the activation of the supposedly dead muscles as being generated by an electrical fluid carried by the still functioning nerves of the frog to the inanimate muscles. Given his background and beliefs, Galvani naturally assumed he had discovered something of the animating, or vital force that was implanted in all creatures by their Creator. However, not everyone agreed with this conclusion. In particular, Alessandro Volta thought that the term 'animal electricity' had a suggestion of superstition and magic and that it was not an explanation of the dramatic phenomenon observed repeatedly by Galvani and his co-workers. For his part, Galvani held that natural philosophers like Volta had no place in moving from the laboratory into God's realm of vitalism, and the nature of life itself. The argument between Galvani and Volta was a microcosm of the larger debate about the place of the Divine in Nature which was animating the European Enlightenment. Galvani spent years repeating his experiment on dead animals and discovered that you did not need a traditional source of static electricity to cause the dead muscle tissue to twitch. A combination of two wires of different metals was sufficient.

The phenomenon observed by Galvani was subsequently named 'galvanism', on the suggestion of his sometime intellectual adversary, Volta. Today, the term galvanism is used only to describe someone who suddenly becomes excited, and it is likely that most people who use this word have no idea of its origin. Although at the beginning of the 19th Century, the observations of Galvani were the source of much discussion, most famously in the novel *Frankenstein, or, The Modern Prometheus* by Mary Shelley, which famously describes further investigations into the principle of animation and vitalism, and the phenomenon of galvanism.

Alessandro Volta was more of a scientist than Galvani. In the late-1770s, Volta had studied the chemistry of gases and was the first person to investigate the origin and chemical composition of natural gas or methane. However, it is for his investigations into the nature of electricity that Volta is most famous; in particular, for a systematic investigation of electrical capacitance. He developed separate means of investigating both the electrical potential applied to the two plates of a capacitor, and the charge residing on the plates. Volta discovered that for a given pair of plates, the potential and the charge are proportional. This relationship is called Volta's Law of Capacitance, and to honour his fundamental work on electrostatics, the unit of electrical potential has been named the volt.

Volta realized from his own studies of Galvani's observations that the frog's leg merely served as both a conductor of electricity (the fluid in the dead muscle tissue is what we would today term an electrolyte) and a detector of the presence of a flowing electric current; all of which mimicked an instantaneous animation. Indeed, Volta realized that the two different metals (the electrodes) used by Galvani, inserted into the fluid of the frog's leg formed an electrical circuit. Volta replaced the frog's leg by paper saturated with another conducting electrolyte, e.g. salt solution, and detected a flow of electricity. In this way, he invented the electrochemical cell, the forerunner of all chemical batteries.

Galvani never perceived of electricity as being separable from biology. He always believed that animal electricity came from the muscle of the animal. Volta, on the other hand, reasoned that animal electricity was merely a physical phenomenon; an electric current coming from the metals (the difference of the electrode potentials), which formed an electrochemical cell, mediated by the fluid in the muscle tissue. There was no reanimation of dead tissue, merely a flow of electrical current from one electrode to the other electrode through the physiological fluid (the electrolyte) in the muscle of the dead frog. But Galvani's ideas did give literature, and the cinema the wonderful Dr Frankenstein and his creature.

In the early-19th Century, electricity, magnetism, and optics were three independent disciplines. However, the situation changed, thanks to one invention and two discoveries. The invention was the electrical battery, a continuous source of electrical current created by Alessandro Volta in about 1800. The two discoveries were: first, the demonstration of magnetic effects caused by the flow of electrical currents; observed by the Danish chemist and physicist Hans Christian Ørsted (1777–1851) and by the French mathematician and one of the creators of the Metric system, André-Marie Ampère (1775–1836) in 1820. And second, the 1831 discovery by the British chemist and natural philosopher Michael Faraday (1791–1867) of the generation of electrical currents from magnetic fields; that is, electromagnetic induction.

In September 1820, Ampére presented his results to the *Académie des sciences*: '*mutual action between currents without the intervention of any magnet*'; that is, two parallel electrical currents attract, or repel each other depending on their polarity, as do permanent magnets. In 1826, he published *Theory of Electrodynamic Phenomena, Uniquely Deduced from Experience*, where he claimed that 'magnetism is merely electricity in motion' and that magnetic phenomena depend only on the existence and motion of electrical charges (see figure 6.2). Thereby setting the stage for Faraday's experiments.

These three contributions form the basis of modern electromagnetism, but it required the insight of the Scot, James Clerk Maxwell (1831–79) to form a coherent single theory. Before Maxwell, electromagnetism still consisted of long lists of observations of, supposedly disparate phenomena; Maxwell demonstrated the single underlying causation. Such a synthesis represents the most profound transformation

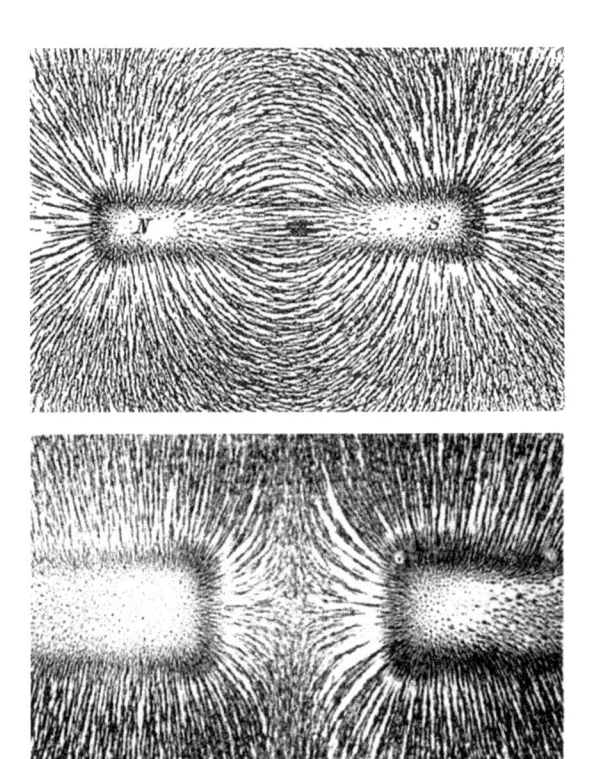

Figure 6.2. Action at a distance. These images give a visualization, in iron fillings, of the force-field extending from permanent magnets. The magnetic field is invisible and extends from the poles of the magnets. In the upper figure, we see the lines of force extending from a single permanent magnet, the North Pole and the South Pole of the magnet are indicated (from: https://en.wikipedia.org/wiki/Magnet#/media/File:Magnet0873.png), and in the lower figure we see the lines of force that represent the repulsive interaction between two like-poles of two permanent magnets (from: https://commons.wikimedia.org/wiki/File:Magnetic_field_of_bar_magnets_repelling.png). The images have been obtained by the author from the Wikimedia website where they are stated to have been released into the public domain. They are included on that basis.

of the fundamentals of physics since Newton, and is one of the greatest of scientific achievements; unifying electrical and magnetic phenomena, and enabling the development of the theory of electromagnetic waves, including light. James Clerk Maxwell published his major work, *A Treatise on Electricity and Magnetism* in 1873; a first step on the great journey to the Theory of Everything. Here, Maxwell rationalised and unified all the then known phenomena involving electricity and magnetism.

6.3 The molecule meme

When we come to the interaction of light; that is, an oscillating electromagnetic field with matter, we need to consider the achievement of James Clerk Maxwell, who between 1861 and 1862 published a set of equations relating electricity and magnetism, and demonstrated that light is an electromagnetic phenomenon. Classically, light scattering arises through secondary radiation from oscillating dipoles induced by the incident electromagnetic wave. The simplest case occurs when the scattering medium is a gas, composed of randomly distributed molecules of dimensions small compared to the wavelength of the light. For a random distribution, the phase relationships between waves scattered from different molecules are uncorrelated in all but the forward direction, so that the total scattered intensity can be calculated directly as the sum of contributions from each molecule; thereby permitting study of the properties of individual molecules. In a real gas, intermolecular forces tend to correlate the phases, but such effects become significant only at elevated pressures, or in the condensed phases. For small molecules, the applied oscillating field can be assumed uniform and the induced dipoles treated as point sources. With these conditions, the incident light induces an oscillating electric dipole proportional to the instantaneous electric field acting on the molecule.

The four equations, commonly called Maxwell's equations, are a set of partial differential equations that, together with the Lorentz force law, underpin all electric, optical, and radio technologies, including, power generation, electric motors, wireless communication, cameras, television, computers, etc. Maxwell's equations describe how electric and magnetic fields are generated by charges and currents. And as molecules are essentially clouds of fast-moving electrons associated with a nuclear framework that deforms at a much slower rate than the distribution of electrons around it, Maxwell's equations are directly applicable to molecules. One important consequence of the equations is that they demonstrate how fluctuating electric and magnetic fields propagate at the speed of light, c.

The equations have two major variants. The microscopic Maxwell equations have universal applicability, but are unwieldy for common calculations. They relate the electric and magnetic fields to total charge and total current. The macroscopic Maxwell equations define two new auxiliary fields that describe the large-scale behaviour of matter without having to consider atomic scale details. However, their use requires experimentally determined parameters for a phenomenological description of the electromagnetic response of materials. In the usual formulation, there are four equations. Two inhomogeneous equations describe how the electric and magnetic fields vary in space due to sources: Gauss's law describes how electric fields emanate from electric charges, and Gauss's law for magnetism describes magnetic fields as closed field lines. Then, there are two homogeneous equations describing how the fields circulate around their respective sources: Ampère's law with Maxwell's addition describes how the magnetic field circulates around electric currents and time-varying electric fields, while Faraday's law describes how the electric field circulates around time-varying magnetic fields. A separate law, the Lorentz force law, describes how the electric and magnetic field act on charged particles and currents.

The formulation of Maxwell's equations depends on the precise definition of the quantities involved. Conventions differ with different unit systems, because various definitions and dimensions are changed by absorbing factors such as the speed of light c. Thus, when manipulating Maxwell's equations, attention must be paid to the unit system being used. The most common form of the equations is based on the SI, but other commonly used conventions require other systems of units; including, Gaussian units based on the cgs (centimetre-gram-second) system, Lorentz–Heaviside units (used mainly in particle physics), and Planck units (used in theoretical physics). The vector calculus formulation of the equations has become standard. It is mathematically more convenient than Maxwell's original formulation, and is due to the English mathematician and electrical engineer Oliver Heaviside (1850–1925).

6.4 Unit conversion in electromagnetism

Gaussian units constitute a metric system (to the base 10) of physical units. This system is the most widely used of the many electromagnetic unit systems based on the cgs system of units. It is also called the Gaussian unit system, Gaussian-cgs units, or simply, cgs units. It is the most widely used, because the units are designed to deal with the small values of quantities arising in molecular physics; as it uses centimetres, grams and seconds.

The most common alternative to Gaussian units is the SI (historically called the mksa system of units for **m**etre–**k**ilogram–**s**econd–**a**mpere). SI units are predominant in most fields, and continue to increase in popularity at the expense of Gaussian units; but their use results in huge values for common quantities such as electric fields and currents. The SI units of electromagnetism were not designed with molecular physics in mind; they are more appropriate for electrical engineering. Conversions between Gaussian units and SI units are not as simple as with mechanical unit conversions; for example, Maxwell's equations need to be adjusted depending on which system of units one choses.

With Maxwell's equations, one difference between Gaussian and SI units are the factors of 4π in the various formulae (see the tables below). SI electromagnetic units are termed 'rationalized', because Maxwell's equations have no explicit factors of 4π in the formulae. On the other hand, the inverse-square force laws, such as Coulomb's law and the Biot–Savart law, do have a factor of 4π attached to the distance dependence, r^2. (This quantity of 4π appears because $4\pi r^2$ is the surface area of the sphere of radius r.) In unrationalized Gaussian units (not in Lorentz–Heaviside units) the situation is reversed. Two of Maxwell's equations have factors of 4π in the formulae, while both of the inverse-square force laws, Coulomb's law and the Biot–Savart law, have no factor of 4π attached to r^2 in the denominator.

Another major difference between Gaussian and SI units is in the definition of the unit of charge. In the SI, a separate base unit (the ampere) is associated with electromagnetic phenomena, with the consequence that something like electrical charge (1 coulomb = 1 ampere × 1 s) is a unique dimension of a physical quantity, and is not expressed purely in terms of the mechanical units (kilogram, metre, and second). On the other hand, in Gaussian units, the unit of electrical charge (the statcoulomb or statC) can be written entirely as a dimensional combination of the mechanical units (gram, centimetre, and second); as for example, 1 statC = 1 $g^{1/2}$ $cm^{3/2}$ s^{-1}; Coulomb's law in Gaussian units is, $F = Q_1 Q_2/r^2$, where F is the repulsive force between two like-electrical charges, Q_1 and Q_2, and r is the distance separating these charges. If Q_1 and Q_2 are expressed in statC and r in cm, then F will be expressed in dyne (a dyne is the unit of force in the cgs system of units; 1 dyn = 1 g cm sec^{-2}; in the SI, force is given in newtons with 1 newton (N) = 1 kg m s^{-2} = 10^5 dyn); the difference of magnitudes is evident, and this problem is aggravated by the use of metres and kilograms in the SI.

By contrast, the same law in SI units is $F = (1/4\pi\varepsilon_0)Q_1Q_2/r^2 = k_eQ_1Q_2/r^2$, where ε_0 is the vacuum permittivity, a quantity with dimension, namely (charge)^2 (time)^2 (mass)$^{-1}$ (length)$^{-3}$, and k_e is Coulomb's constant. Without ε_0, the two sides would not have consistent dimensions in the SI; in fact, the quantity ε_0 does not exist in Gaussian units. This is an example of how some dimensional physical constants can be eliminated from the expressions of physical laws simply by the judicious choice of units. In the SI, $1/\varepsilon_0$ converts or scales flux density, **D**, to electric field, **E** (the latter has dimension of force per charge), while in rationalized Gaussian units, flux density is the same as electric field in free-space.

Since the unit of charge is built out of mechanical units (mass, length, and time), the relation between mechanical units and electromagnetic phenomena is clearer in Gaussian units than in the SI. In particular, in Gaussian units, the speed of light c shows up directly in electromagnetic formulae such as Maxwell's equations, whereas in the SI it only shows up implicitly via the relation $\mu_0\varepsilon_0 = 1/c^2$. In Gaussian units, unlike SI units, the electric field **E** and the magnetic field **B** have the same dimensions. This amounts to a factor of c difference between how **B** is defined in the two-unit systems (the same factor applies to other magnetic quantities such as **H** and **M**); for example, in a planar light-wave in vacuum, $|\mathbf{E}(r, t)| = |\mathbf{B}(r, t)|$ in Gaussian units, while $|\mathbf{E}(r, t)| = c|\mathbf{B}(r, t)|$ in SI units. We will see later that the recent redefinitions of the SI have profoundly changed the manner in which we consider the permittivity (ε_0) and permeability (μ_0) of free space.

The four Maxwell equations describe the properties of electric and magnetic fields in various media. The four equations relate the six properties listed in table 6.1.

As mentioned above, the electric displacement and the magnetic flux density are related to the strengths of the electric and magnetic fields by the polarization **P** and the magnetization **M**: $\mathbf{D} = \varepsilon_0\mathbf{E} + \mathbf{P}$ and $\mathbf{B} = \mu_0\mathbf{H} + \mu_0\mathbf{M}$.

The four equations due to Maxwell (as modified by Heaviside) are given in table 6.2, which also includes the Lorentz force law.

Table 6.1. The six quantities described, and related by Maxwell's equations.

	Property	SI units	Gaussian units
E	Electric field strength	V m^{-1}	statvolt cm^{-1}
D	Electric displacement	C m^{-2}	statvolt cm^{-1}
ρ	Charge density (a scalar)	C m^{-3}	statcoul cm^{-3}
H	Magnetic field strength	A m^{-1}	oersted
B	Magnetic flux density	Tesla or Wb m^{-2}	gauss
J	Current density	A m^{-2}	statamp cm^{-2}

Table 6.2. Maxwell's equations and the Lorentz force law in SI and Gaussian units.

Name	SI units	Gaussian units
(a) Gauss' law	$\nabla . \mathbf{D} = \rho$	$\nabla . \mathbf{D} = 4\pi\rho$
(b) Gauss' law for magnetism	$\nabla . \mathbf{B} = 0$	$\nabla . \mathbf{B} = 0$
(c) Maxwell–Faraday law (Faraday's law of induction)	$\nabla \char`^ \mathbf{E} = -\partial \mathbf{B}/\partial t$	$\nabla \char`^ \mathbf{E} = -(1/c)\ \partial \mathbf{B}/\partial t$
(d) Ampère–Maxwell equation	$\nabla \char`^ \mathbf{H} = \mathbf{J} + \partial \mathbf{D}/\partial t$	$\nabla \char`^ \mathbf{H} = (4\pi/c)\mathbf{J} + (1/c)\partial \mathbf{D}/\partial t$
Lorentz force law	$F = q(\mathbf{E} + \mathbf{v} \times \mathbf{B})$	$F = q(\mathbf{E} + (1/c)\ \mathbf{v} \times \mathbf{B})$

Over the first 30 years of the 19th Century, phenomena involving electricity and magnetism (electromagnetism) would remain a collection of disparate observations, and it was not until the second half of that century that a single synthesis of the various observations involving electric and magnetic phenomena was made by the Scottish mathematical physicist James Clerk Maxwell. Indeed, it would be true to say that the development of Maxwell's equations of electromagnetism created a coherent model for these phenomena in the same way that the formulation of the First and Second Laws of Thermodynamics created a coherent model of energy and temperature.

Further reading

[1] Clemmow P C 1973 *An Introduction to Electromagnetic Theory* (Cambridge: Cambridge University Press)

Chapter 7

Measurement in the age of scientific certainty

For the ordinary people of France, the Metric System represented not just stand-ardization of the world but a revolutionary rupture in French society. The old French units of measurement, like all such systems of weights and measures, were inconsistent in part because they were meant to be variable; they had evolved that way. In mediaeval France, two old land measures, the *homme* (the man) and the *journée* (the day), for example, were based not on fixed dimensions of land, but on the amount of work a labourer could perform in a day. The size of this quantity of work would vary from one plot of land to a neighbouring plot of land, from one worker to another worker; it would vary due to the nature of the soil in which the labourer was working, and whether or not the land being worked was hilly or flat; and the payment was always subject to negotiation. However, a square kilometre of land, unlike a *journée*, was unchanging and impersonal, and the common people of France resisted the change to the new metric units for decades. The Metric System was seen as another stage in the evolution of modern economic man, which replaced the older social cohesion, and was feared by the working man.

Let us now consider how countries other than France were dealing with the need for more reliable systems of weights and measures at the beginning of the 19th Century. In America, we saw how the first president, George Washington, had kept telling the US Congress to do something about weights and measures, but his prodding had only produced an ineffectual report written by the Secretary of State, Thomas Jefferson. The problem of weights and measures had, however, not gone away. In 1816, President James Madison said in his address to Congress that 'Congress will call to mind that no adequate provision has yet been made for the uniformity of weights and measures contemplated by the Constitution. The great utility of a standard fixed in nature and founded on the easy rule of decimal proportions is sufficiently obvious.'

In 1817, the Congress got around to requesting the then Secretary of State, John Quincy Adams, to recommend a system of weights and measures for use in the USA. Following a detailed four-year investigation, John Quincy Adams submitted a comprehensive report to the US Congress. This report dealt with the modernization of the measurement system then in use in the USA, and included some thoughts on 'the metric question'. Adams commented 'Weights and Measures may be ranked among the necessaries of life to every individual of human society. They enter into the economical arrangements and daily concerns of every family. They are necessary to every occupation of human industry; to the distribution and security of every species of property; to every transaction of trade and commerce; to the labors of the husbandman; to the ingenuity of the artificer; to the studies of the philosopher; to the researches of the antiquarian; to the navigation of the mariner; and the marches of the soldier; to all the exchanges of peace, and all the operations of war. The knowledge of them, as in established use, is among the first elements of education, and is often learned by those who learn nothing else, not even to read and write. This knowledge is riveted in the memory by the habitual application of it to the employments of men throughout life.'

Not only was Adams, putting down one of the clearest statements of why a nation; particularly, a new nation should maintain a reliable system of weights and measures, but he was also pointing out why a system of weights and measures is so fundamental to the life of a nation, and to all of the citizens of that nation.

When Adams presented his *Report Upon Weights and Measures* to the US Congress, he recommended that the Congress give consideration to the Metric System. Specifically, he commented, 'The [*metric*] system approaches to the ideal perfection of "uniformity." [*It*] will shed unfading glory upon the age .... Considered merely as a labor-saving machine, it is a new power, offered to man, incomparably greater than that which he has acquired by the new agency which he has given to steam. It is in design the greatest invention since that of printing.' However, Adams went on to recommend that no immediate, fundamental change to the system of weights and measures be made. Adams considered that it would be premature for the United States to adopt the Metric System, before it had demonstrated its utility and become a practical success in other parts of the world. It is likely that Adams was influenced by the decision of the Emperor Napoléon to return to the *mesures usuelles* in 1812. One could say that had it not been for Napoléon's populist sentiments in limiting the use of the Metric System in France, the USA could well have adopted metric measures in the early-19th Century.

In Britain, things were very different; The Weights and Measures Act of 1824 established new units of measurement for Britain and her colonies (but by this time, America was no longer a colony), thereby creating the so-called British Imperial units. The three basic units were the *avoirdupois* pound, the yard, and the second. The imperial unit of length was defined as the length of the prototype of the imperial yard which had been made in 1760 and was stored at the Palace of Westminster. All measurements of length were to be based on this artefact of the prototype yard. The imperial unit of capacity, the gallon, was defined as the volume of ten *avoirdupois* pounds of distilled water, weighed in air against brass weights at a temperature of 62 °F, and at an atmospheric pressure of 30 inches of mercury.

This Act of Parliament demonstrated that the UK was a rapidly industrializing nation and that Parliament had decided to intervene and define a standard system of units and measurements which were to be used throughout the UK. This parliamentary intervention in defining the technical details of the rapidly expanding industry of Great Britain also meant that Britain's trading partners around the world would also have to adopt British Imperial units if they wished to trade at significant levels with the UK. Consequently, as far as the British were concerned, the Metric System had come along too late to be of any use to them, as British industry was already well established, and it was tooled in yards, inches and pounds.

Even as the politicians of France, and of some of her Continental trading partners finally accepted the Metric System, the new 'scientists' of the 19th Century were developing the ideas of their predecessors. During the early-part of the 19th Century, science had finally begun to present a coherent picture of the phenomena of electricity and magnetism, and of the concept and nature of energy. And it was immediately apparent that new units of measurement would be required to fully comprehend and quantify these new phenomena.

During the early-1850s, German scientists Carl Friederich Gauss (1774–1855) and Wilhelm Weber (1804–91) worked to develop a system of units for electricity and magnetism. Their system was extended and promoted in Britain by the Scottish mathematician James Clerk Maxwell and the Irish physicist William Thomson (1824–1907), who later became Lord Kelvin, under the auspices of the British Association for the Advancement of Science. Maxwell and Thomson helped to formulate the requirements for a coherent system of units with base units, and derived units for a whole range of newly-discovered phenomena. They were creating an early form of the language of science needed to fully describe all nature, that is, to comprehend newly discovered phenomena.

It was Kelvin, one of the greatest physicists of the 19th Century, who played an early role in the creation of the International System of Units (SI). He was known to have called the British customary (Imperial) system of weights and measures 'barbarous', and he had good reason. In a lecture demonstration with a muzzle-loading rifle, confusion on the part of a student between the *avoirdupois* dram (about 1.8 g) and the apothecary's dram (about 3.9 g) caused the student to put twice as much gunpowder as he should have into Kelvin's rifle. And it was only Kelvin's attention to detail, which led him to check the amount with the student before the probably fatal demonstration.

The rapid industrialization of the first half of the 19th Century was accompanied by and could not have occurred without a corresponding advance in the design and manufacture of machine tools. That is, machines for measuring distance in a factory workshop and not a laboratory; machines that allowed standardization of screw threads, and for measuring flat surfaces and straight edges. All of which are essential for precision manufacturing on a large scale. By the middle of the 19th Century, engineering metrology had reached an appropriate level of precision, and the stage was thus set for the birth of the modern industrial world.

While these advances were being made in engineering metrology and industrial production, there had been no changes in the metric standards of length and mass established in 1795. The first Great Exhibition, of 1851 in London, brought together manufactured products from all over the world and demonstrated the great advances that had been made in mechanical engineering; particularly, in Britain and in mass production, notably in the USA. It was clear that although great advances were being made, there was still the problem of the different systems of units of measurement—not only in measurements of length and mass.

At the second Great Exhibition, of 1855 in Paris, formal moves began to establish worldwide agreement on units of measurement. The *Commissaires* and members of the jury judging the exhibits in Paris made a formal request at the closure of the Exhibition to governments for the creation of a worldwide system of weights and measures based on the decimal metric system. This plea was supported by a request from the Society of Arts and Manufacturing in London to the British government for the introduction of the Metric System into Britain and the British Empire. However, it was only in 1864 that the use of the Metric System became legally possible in Britain.

At the 1867 Great Exhibition, again held in Paris, a Committee for Weights and Measures and Money was created. In that same year, the Academy of Science of Saint Petersburg and the newly formed International Association for Geodesy (geodesy is the study of how the rotation of the Earth impacts upon the concept of time as measured by man) both asked governments to work towards establishing a single international system of weights and measures. The French *Bureau de Longitude* transmitted these requests to the French Government, and then with representatives of the *Académie des sciences* made a formal request to the French Government for the creation of an international commission to oversee the construction of new international metric standards, with a view to distributing these new standards to all major industrial nations, and thereby making the Metric System truly international.

Such an international commission was created in 1870 by the government of the French Emperor Napoléon III (a nephew of Napoléon Bonaparte), who was the Head of State of the French Second Empire, which lasted from 1852 to 1870. Unfortunately, the work of the new commission was interrupted by the disastrous defeat of France in the Franco–Prussian War of 1870–1871. In 1872, the new French Third Republic (Napoléon III had been forcibly retired to England) convened an international commission (the International Commission of the Metre) to meet in Paris. This new international commission became the forerunner of the General Conference of Weights and Measures (*Conférence générale des poids et mesures*), which still meets every four years in Paris. Representatives of both the UK and the USA attended the meetings of the International Commission of the Metre.

7.1 The Convention du mètre

After a great deal of negotiation, the meetings of the International Commission of the Metre eventually led to the signing of the Metre Convention on 20 May 1875 (see figures 7.1 and 7.2). This Convention represents the biggest single step in the international promotion of the Metric System. The Metre Convention (*Convention du mètre*) is a diplomatic treaty signed by most of the nations who sent representatives to the meetings of the International Commission of the Metre.

Figure 7.1. Medal cast to commemorate the work of the International Commission of the Metre, which led to the *Convention du mètre* signed in Paris on 20 May 1875. The face of the medal is decorated with allegorical figures; Science holding the new metric standard, surrounded by the figures of Europe, America and Asia. The Latin text reads 'Populorum concordiae sacrum' (Testimony of the peoples' agreement. Paris 1872). (This image is reproduced with the permission of the BIPM, which retains full international copyright.)

The *Convention du mètre* established a permanent structure of technical expertise, within which governments could discuss matters relating to units and standards of measurement. It set in place the manner in which international activities related to weights and measures should be financed and managed, and of particular importance for the remainder of this book, how existing systems of weights and measures may be renewed and redefined. The main decision was to set up three separate, but interacting bodies: the *Conférence générale des poids et mesures* (CGPM), in English, the General Conference on Weights and Measures, which would meet every four years in Paris, the *Comité international des poids et mesures* (CIPM), in English, the International Committee for Weights and Measures, which would meet annually, and the *Bureau international des poids et mesures* (BIPM), in English, the International Office of Weights and Measures, which would be a permanent laboratory based near Paris. The BIPM operates under the supervision of the CIPM, which itself comes under the authority of the CGPM. The *Convention du mètre*, thus gives authority to the CGPM, the CIPM and BIPM to act at an international level in the definition, promotion and perfecting of measurement standards. This is the organizational structure that still exists today to resolve major problems of metrology, and to maintain and develop the modern scientific version of the Metric System.

7.2 Conférence générale des poids et mesures (CGPM)

The CGPM consists of delegates from the Member States (signatories) of the Metre Convention who meet every four years in Paris. It is a formal diplomatic organization responsible for the maintenance and evolution of an international system of units of measurement. Consequently, the CGPM must take note of the latest scientific and technological advances.

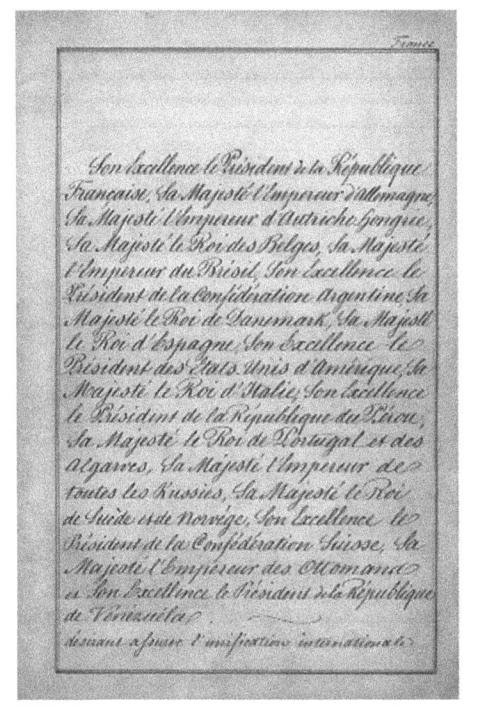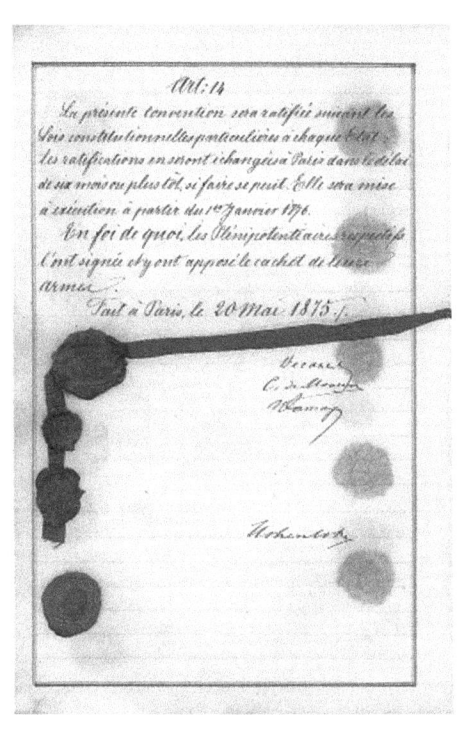

Figure 7.2. Parts of the *Convention du mètre* or Metre Convention signed in Paris on 20 May 1875. The left-hand page lists the countries about to sign the Convention, and the right-hand page is the first page of signatures and seals. The first signatures are those of the Ministers Plenipotentiary for France, which is the guardian of the treaty and why French is the only official language of all matters relating to the Metric System (representatives of the coalition government of Louis Buffet), Louis-Charles-Élie-Amanien Decazes de Glücksbierg, 2nd Duke Decazes and 2nd Duke of Glücksbierg (1819–86) as Minister for Foreign Affairs, the *vicomte* Marie-Camille-Alfred de Meaux the Minister for Agriculture and Commerce and *Monsieur* Jean Baptiste André Dumas as *secrétaire perpétuel* of the *Académie des sciences* and one of the leading physiologists of the day. (This image is reproduced with the permission of the BIPM, which retains full international copyright. The Treaty itself is kept at the French Ministry of Foreign Affairs at the *Quai d'Orsay*.)

7.3 Comité international des poids et mesures (CIPM)

The CIPM consists of 18 individuals each belonging to a different Member State of the Metre Convention, and it meets every year at the BIPM near Paris. The CIPM reports to the CGPM on the work that the BIPM has accomplished in the previous four years and indicates the direction it intends to take in the next four years.

7.4 Bureau international des poids et mesures (BIPM)

Since the signing of the Metre Convention, all official copies of the original prototype of the kilogram (the International Prototype of the Kilogram, figure 7.4, which until May 2019 had been the SI base unit of mass, and which is still preserved at the BIPM, as is the original metre artefact) have been manufactured at the BIPM (see figure 7.3).

Figure 7.3. The seal of the BIPM, representing an allegory of Science holding in her hands the new metre standard with its decimal divisions. Mercury or Hermes, the god of commerce, invention and weights and measures wearing a winged cap is an allegory of industry, he carries his herald's staff and, as befits the messenger of the gods, sits upon a map. The other feminine figure in the seal carries the symbols of industry, the mallet and a gear wheel. The seal carries the inscription in Greek *Metro Kro* or 'use the measure'. It is suggested that the laurel decorating the ends of the metre represents the triumph of the work of those who made the new metric prototypes. (This image is reproduced with the permission of the BIPM, which retains full international copyright.).

Figure 7.4. The International Prototype of the Kilogram under its three protective glass bell-jars. This object was the pivot for the world's artefact system of mass metrology, which existed between 1889 and 2019. This small object (the density of the alloy from which it is made, means that a kilogram occupies a smallish volume) was once one of the most derided objects in science. The mass standard is under three nested glass bell-jars. The IPK has been conserved at the BIPM since 1889, when it was sanctioned by the 1st CGPM. It is of cylindrical form, with diameter and height of about 39 mm, and is made of an alloy of 90% platinum and 10% iridium. Initially, the IPK had two official copies; over the years, one official copy has been replaced and four others have been added, so that there are now six official copies. Access to the IPK and its official copies is under the strict supervision of the CIPM. (Copyright for this image is with the author.)

The BIPM has responsibility for the worldwide uniformity of measurement, and for ensuring a single, coherent system of measurements traceable to the International System of Units (SI). Where traceability to the SI is not yet possible, for example, in some areas of biology, biotechnology and medicine, the BIPM seeks to identify routes of traceability to other internationally accepted measurement standards.

Prior to May 2019, the BIPM fulfilled its role in international metrology by the direct dissemination of SI units, as in the case of mass (by manufacturing and distributing kilogram artefacts; figure 7.4 shows the International Prototype of the Kilogram). In addition, the BIPM is responsible for the dissemination of atomic time, by coordinating atomic time scales by regular calculations involving the data from atomic clocks from around the world. The BIPM also organizes international comparisons to validate the consistency of national standards, as in electricity, ionizing radiation and chemistry.

The justification for the work undertaken by the BIPM stems from its unique position. It offers a neutral, independent position working for all the Member States of the Metre Convention. It is not an institution created and supported by one nation; and although it is located in France, it is in fact on a neutral, extra-territorial enclave. In this way, the BIPM attempts to continue the non-national dreams of the founders of the Metric System; 'For all time. For all people'[1].

The first CGPM of 1889 approved new international prototypes for the metre and the kilogram, together with the astronomical second as the unit of time. The metre, kilogram, and second now formed a coherent system of scientific units (metre-kilogram-second) which therefore became known as the mks system of units. The first CGPM also formally required that the international prototypes of the metre and the kilogram be deposited at the *Pavillon de Breteuil* in Sèvres, the home of the BIPM for the use of all nations.

Before continuing, we need to consider the distinction between base units and metrological or measurement standards. A base unit is fixed by its definition (from 1795 to 1960, the metre was defined as being one ten millionth of the Earth's quadrant), and was independent of physical conditions such as temperature. By contrast, a metrological prototype or measurement standard is a physical object that can be studied in a laboratory; it is a physical realization of a base unit, and realizes that base unit only under certain physical conditions. For example, the metre is a unit, while a metal bar of length one metre is a measurement standard. One metre is the same length regardless of temperature, but a metal bar will be precisely one metre in length, only at a certain precise temperature due to expansion. Since May 2019, however, the system of base units of the SI has changed dramatically with the creation of the Quantum-SI. There are no longer any artefacts in the SI; the quantities that make up the SI are defined in terms of constants of Nature (see chapters 11 and 12)

---

[1] Full details of the origin and work of the BIPM and of the Metre Convention may be found on the BIPM's website: www.bipm.org

Each member state of the Metre Convention received copies of the metric measurement standards of length and mass with calibration certificates which related them to the international prototypes. These copies of the international standards were sent to the national standards laboratory of the member states. As copies, they were accurate but not identical to the prototypes kept at the BIPM. As a signatory of the Metre Convention, the USA received a prototype metre and kilogram as measurement standards; they were received by President Benjamin Harrison at the White House and were then placed in the vault of the Treasury Department. Britain signed the Metre Convention in 1884, and it was in November 1889 that copies of the international prototypes of the metre and of the kilogram were received in London by the Conservative government of Robert Cecil, the 3rd Marquis of Salisbury.

A serious problem with any prototype standard, or artefact, even if it is the International Prototype of the Kilogram (figure 7.4), or of the Metre is that there is no method to detect a change in its value due to ageing or misuse. Consequently, it was not possible to state the accuracy or stability of the prototype copies of the metre and kilogram standards sent to the member states, although calibration uncertainties of the copies of the metre and kilogram could be assigned. Indeed, it was the question of the long-term stability of artefact standards that finally lead to the final abandonment of artefact-based definitions of base quantities in 2019.

In 1893, the US Congress defeated yet another Bill for the adoption of the Metric System by the USA. However, the prototype metre and a prototype kilogram, which had been received from the BIPM, were declared the nation's fundamental standards by an administrative action of Thomas C Mendenhall, Superintendent of Weights and Measures. The US Secretary of the Treasury then sanctioned this decision and the metric prototypes were legally declared to be the fundamental standards of length and mass for the USA. From 1893, following the Mendenhall Order, the USA has defined the old measures, such as the yard and the pound in terms of the Metric System. The USA yard was defined in the Mendenhall Order as being equal to 0.914 401 83 m, and one inch was equal to 25.400 050 8 mm.

In 1894, the US Congress began yet another study into the adoption of the Metric System by the USA. The Congress passed a Bill to adopt the Metric System; then sent this Bill to the Committee on Coinage, Weights and Measures for further examination. Although attempts have been made to finally enact this Bill, it remains in the limbo of this Congressional Committee.

The slow acceptance of the Metric System in the USA was not limited to Federal legislation. In preparation for its entry into the Union, in January 1896 as the 45th State of the Union the politicians of Utah were writing the State Constitution. Interestingly, the founders of Utah proposed, in Article X, section 11 of the Constitution of the State of Utah, that, 'The Metric System shall be taught in the public schools of the State'. Unfortunately, this section was later repealed.

In 1896, after a comprehensive debate the British Parliament passed the Weights and Measures (Metric System) Act: *An act to legalise the use of weights and measures of the metric system.* This Act legalized the Metric System for all purposes, including everyday commercial transactions, but did not make it compulsory. This Act standardized the imperial gallon by defining it in terms of the cubic decimetre or litre. The Act further permitted the use of metric weights and measures in trade, and required the Board of Trade to include metric denominations among its standards. The imperial yard was measured against the international standard metre and found to be: one yard = 0.914 399 m, and one inch was found to be equal to 25.399 972 mm. This is the Act of Parliament that most authorities cite as the beginning of metrication in the UK.

Further reading

[1] Marquet L, Le Bouch A and Roussel Y 1996 *Le Système Métrique, Hier et Aujourd'hui* (Editions A.D.C.S.)
[2] Moreau H 1975 *Le Système Métrique* (Paris: Chiron)
[3] Quinn T 2011 *From Artefacts to Atoms: The BIPM and the Search for Ultimate Measurement Standards* (Oxford: Oxford University Press)

Chapter 8

A true universal language: the SI

8.1 Even scientists cannot always agree on units

We saw earlier how the derived units of the SI are constructed algebraically from the base units of the SI, using the law of physics. However, sometimes a choice arises between different equations governing the phenomenon of interest, and the manner in which a derived unit may be constructed. An important example of such possible confusion occurs in the manner of defining the quantities describing electricity and magnetism.

The theory of electricity and magnetism (electromagnetism) was developed during the early-19th Century as discoveries were made by Hans Christian Ørsted, André-Marie Ampère, and Michael Faraday. The possibility of defining magnetic phenomena in terms of mechanical units such as length, mass and time; that is, creating a coherent universal system of units was first proposed in 1833 by the German mathematician Carl Frederick Gauss (1777–1855). His analysis was extended to cover electrical phenomena by Wilhelm Weber (1804–91), who in 1851 proposed a method by which a complete set of units could be constructed which incorporated electromagnetism into the existing Metric System.

Systems of units and the ability to convert between different systems of units; for example, to convert from British customary system of units to SI units is something that is no longer taught to science students. This is a great shame, as the different systems of units are only dialects of the single universal language of science, and an inability to communicate with people speaking these different dialects can limit a scientist's world-view. Converting between mechanical quantities expressed in British customary units and those expressed in the SI is relatively easy. However, you cannot go far in electromagnetism before you encounter serious problems in converting between systems of units.

Scientists have spent almost a century disagreeing about the units for electromagnetism. But even if every scientist were suddenly to adopt SI units for electromagnetism, the need for familiarity with the older cgs (**c**entimetre-**g**ram-**s**econd) system of units would still be needed so as to facilitate reading of the vast and published literature in this area of science since the early-19th Century.

The problem begins with deciding what exactly constitutes an electric or a magnetic field. In 1861, a committee of the British Association for the Advancement of Science (BAAS) that included William Thomson (later Lord Kelvin), James Clerk Maxwell, and James Prescott Joule undertook a comprehensive study of electrical measurements. This committee introduced the concept of a comprehensive system of units for electromagnetism. It was discovered that only four simple equations were required to define and couple the units of electrical charge Q, electric current I, voltage or potential difference V, and electrical resistance R. These simple equations are:

• **either** Coulomb's force law (named after Charles-Augustin de Coulomb,) for charges, which tells us that like charges repel each other and unlike charges attract each other, **or** Ampère's force law (named after André-Marie Ampère) for currents (that a repulsive force will develop between two neighbouring parallel wires through which electrical currents are passing in the same direction). Ampère's law was the basis for the old definition for the base unit of electric current (see table 11.1).

• The relationship between electric charge and electric current that tells us that the charge (Q) passed through a conductor is equal to the electrical current (I) passed multiplied by the time for which the current flows (T); that is, $Q = IT$ (this is the basis of the new definition of the ampere, see table 11.1).

The two additional equations needed to define practical electromagnetism are:

• The well-known law due to Georg Simon Ohm (1789–1854) that an electric current (I) passing through a wire with a certain electrical resistance (R) will develop a voltage (V), such that $V = IR$ (Ohm's law).

• The equation for electrical work; that the work (W) done or energy expended when an electrical current (I) passes through a wire of resistance (R) for a time T is given by $W = I^2RT$.

A fundamental principle of the new system of units desired by those 19th Century scientists, seeking to incorporate electromagnetism into the existing systems of mechanical units was that the new system of units should be coherent. That is, the system must be based upon certain base units for fundamental quantities such as length, mass, and time. Then derived units could be constructed for a vast number of other scientific quantities as products or quotients by means of quantity calculus, but without requiring additional numerical factors. The metre, gram, and second were originally selected as base units by the BAAS committee in 1861. Consequently, this was not a system of units based on the Metric System, which has the kilogram not the gram as the unit of mass, and so there is a complicating numerical factor of 10^3. (This point is discussed in chapter 7.)

Because one could use either the force law due to Coulomb or the force law due to Ampère to define electromagnetic quantities, two parallel systems of units arose: the electrostatic and electromagnetic subsystems of the cgs system of units, depending on whether the law of force for electric charges (Coulomb's Law) or the law for electric currents (Ampère's Law) was used as the origin of these forces. The BAAS also organized research on electrical standards for use by the new electricity industry, and issued a wire resistance measurement standard, the 'British Association unit', which later became known as the 'ohm'.

In 1888, the German physicist Heinrich Hertz (1857–94) verified Maxwell's predictions of the nature of electromagnetic radiation and light. His experiments greatly extended the range of frequencies over which Maxwell's equations were shown to be valid and he invented radio (by accident). In addition, Hertz combined the electrostatic and electromagnetic subsystems of the cgs system of units into a single system where the various terms were related by the speed of light c; he called this new system the Gaussian system of units in honour of Karl Friedrich Gauss.

The recommendations of the BAAS of 1873 were adopted by the first International Electrical Congress in Paris in 1881. Five practical electrical units were defined: the ohm (named for George Ohm), the farad (named for Michael Faraday), the volt (named for Alessandro Volta), the ampere (named for André-Marie Ampère), and the coulomb (named for Charles-Augustin de Coulomb). In 1889, the second International Electrical Congress added to this list of units: the joule to name a quantity of energy, in honour of James Prescott Joule, the watt to name a quantity of power, in honour of James Watt, and a unit of electromagnetic inductance, later given the name henry in honour of the American physicist Joseph Henry (1797–1878).

Table 8.1 gives several derived units used in the SI, with their special names and symbols. One can see that these derived units are more complex than those derived units given in chapter 5. However, science had advanced rapidly in the 19th Century, and we see here the increasing need for units to describe newly discovered phenomena; a process which is continuing today with modern discoveries.

In 1901, an Italian electrical engineer Giovanni Giorgi (1871–1950) demonstrated that the BAAS electrical units and the mks (**m**etre-**k**ilogram-**s**econd) mechanical units (derived from the Metric System), introduced by the 1st CGPM of 1889, could be incorporated into a single coherent system of units by selecting the metre, kilogram, and second as the base units for mechanical quantities, and adding a base unit of an electrical nature, and, finally, by giving a physical character to the space

Table 8.1. Derived units involving electromagnetic units (A is the base unit of electricity, the ampere—see figure 8.1).

Derived quantity	Derived unit and symbol	Expressed in terms of other SI units	Expressed in terms of SI base units
Electric charge	coulomb (C)		s A
Electrical potential	volt (V)	W A^{-1}	m^2 kg s^{-3} A^{-1}
Capacitance	farad (F)	C V^{-1}	m^{-2} kg^{-1} s^4 A^2
Resistance	ohm (Ω)	V A^{-1}	m^2 kg s^{-3} A^{-2}
Magnetic flux	weber (Wb)	V s	m^2 kg s^{-2} A^{-1}
Magnetic flux density	tesla (T)	Wb m^{-2}	kg s^{-2} A^{-1}

through which the electric or magnetic field propagated. This latter point required the assignment of physical dimensions to the permeability of free space (this is a property of the vacuum) through which electric and magnetic fields propagate.

With the need for further international cooperation on units of electromagnetism, the 6th CGPM of 1921 amended the Metre Convention of 1875 so that units of electromagnetism would be regulated by the institutions created by the Metre Convention. By the 8th CGPM in 1933, there was an international desire to assist the electricity industry by creating an absolute set of electrical units based on Giorgi's proposals, with the practical electrical units incorporated into a comprehensive mks system of units.

In September 1935, the world of measurement science accepted that the ampere should be chosen as the base unit for electricity, and that it be defined in terms of the force per unit length between two long parallel wires (see table 11.1). By 1946, international organizations with an interest in the international coordination of the physical sciences accepted Giorgi's proposal for a four-dimensional system of units based on the **metre-kilogram-second-a**mpere (mksa system of units), to take effect on 1 January 1948.

Following the mechanism for international metrological collaboration created under the Metre Convention, the CGPM instructed the CIPM to investigate the creation of a complete set of units of measurement. In addition, the CIPM was instructed to investigate the opinion prevailing in the member states of the Metre Convention on the creation of a modern system of units based on the Metric System. These moves meant that the mksa system quickly evolved into the Système international d'unités, or SI (International System of Units) of 1960.

In summary, the various stages of the evolution of the systems of units which we today call the International System of Units (represented in its present form in figure 8.1), and which derived from the Metric System of 1795 are as follows:

- 1668: Bishop John Wilkins' proposal for a universal measure.
- 1795: The Metric System created in Revolutionary France.
- The cgs metric system of about 1872, which then split into an electrostatic-cgs system and a magnetic-cgs system of units, depending upon the choice of Ampère's Law or Coulomb's Law to define force.
- 1875: The Metric System was proposed to all nations through the Metre Convention.
- 1901: Giorgi's mksa proposal for a coherent system of metric units.
- 1948: Giorgi's mksa system was formally adopted, and the other metric systems of units were abandoned.
- 1960: Creation of the SI.
- 2019: Redefinition of the base units of the SI, and the creation of the Quantum-SI (see figure 8.1)[1]. The changes in the definitions of the base units are given in table 11.1.

---

[1] As the Quantum-SI is so very new, for completeness we have included figure 8.2 which gives the interconnectivity in the old-SI (no longer used). Comparing the interconnectivity displayed in figures 8.1 and 8.2 provides an insightful overview of the evolution of physics over second half of the last century.

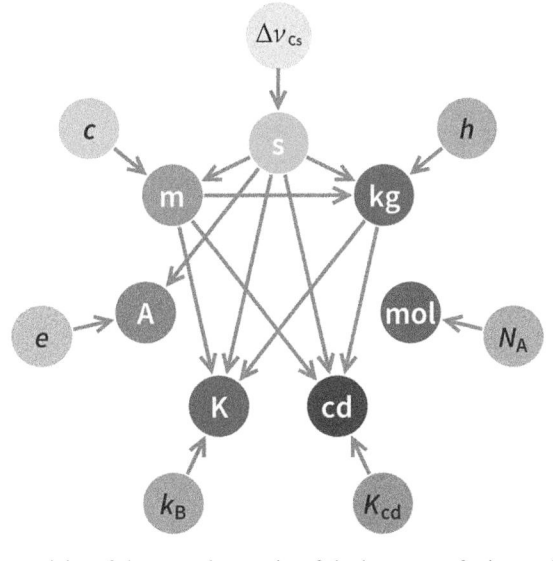

Figure 8.1. The interconnectivity of the seven base units of the language of science; that is, the International System of Units (SI). The base quantities of the new Quantum-SI are represented by seven, centrally located small circles: time (s), mass (kg), length (m), amount of electricity (A), thermodynamic temperature (K), light intensity (cd), and amount of substance (mol). The outer ring of coloured circles represents the seven constants of Nature upon which each of the base quantities depend (see chapters 11 and 12 for a full explanation). The arrows indicate the complex interdependence of these seven, supposedly independent, base quantities. For example, the metre is defined by the speed of light (c); the second is defined by the frequency of a hyperfine transition in a caesium atomic clock (Δv_{Cs}); the kelvin is defined by the Boltzmann constant (k_B); the kilogram, which was defined by the artefact that is the International Prototype of the Kilogram of IPK (see figure 8.2) is now defined by the Planck constant (h); the mole is defined by Avogadro's number (N_A); and the ampere is defined by the charge of the electron (e). This web of interconnectedness was different in the old SI (see chapters 11 and 12 for details). (Image from https://en.wikipedia.org/wiki/2019_redefinition_of_the_SI_base_units#/media/File:Unit_relations_in_the_new_SI.svg where it was made available by Emilio Episanty under a CC BY-SA 4.0 licence, and is reproduced by kind permission of Dr Emilio Pisanty.)

It can be seen from figure 8.1 that, in the Quantum-SI, it is only the base units of time and of quantity of substance that are truly independent of the definitions of any other of the base units, and that the base unit of time is implicit in the definition of the base units of length (the metre) and mass (the kilogram). The candela, the base unit of light intensity, can be seen to be dependent upon the definitions of the base units of length, mass, and time. Light intensity is a change of energy ($M\,L^2T^{-2}$) with time and so is based on the base units of length, mass, and time. Consequently, any future change to the definitions of the base units of length, mass, or time will necessitate a change in the definition of the base unit of light intensity. Previously, because of the manner in which the ampere was defined (see table 11.1), as a force ($M\,L^2T^{-2}$) between two wires carrying an electric current, the definition of the ampere depended upon the definition of length, mass, and time. This is no longer the case, but the ampere is still dependent upon the base unit of time, as it is defined as the number of electrons flowing in a conductor in a set period of time. Thus, the

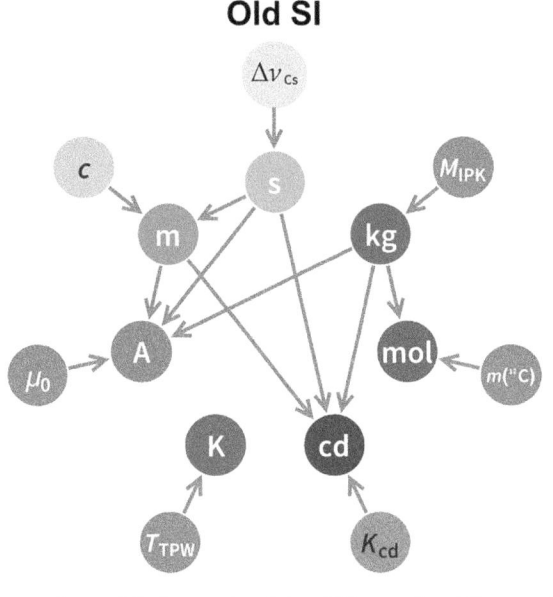

Figure 8.2. The interconnectedness of the base units of the old International System of Units (SI), as it existed between 1983 and the May 2019 redefinition. One can readily see the dependence of base unit definitions on other base units; for example, the definition of the second is implicit in the definitions of the metre, the ampere and the candela. In the Quantum-SI, the definition of the second is implicit in the definitions of all the other base units, with the exception of the mole. In the old-SI, the kilogram was defined by an artefact (see figure X), but now it is defined by the value of the Planck constant. In the old-SI, the ampere was defined as a force between two current-carrying wires; today, it is defined by the flow of a certain number of electrons. See table X for a comparison of the old and the new definitions. (Image from https://en.wikipedia.org/wiki/2019_redefinition_of_the_SI_base_units#/media/File:Unit_relations_in_the_old_SI.svg, where it was made available by Emilio Episanty under a CC BY-SA 4.0 licence and is reproduced by kind permission of Dr Emilio Pisanty.)

seven base units are not completely independent, but are in fact strongly coupled together. But that coupling is different in the Quantum-SI (figure 8.1) and in the old-SI (figure 8.2); compare the arrows coupling the inner-ring of coloured circles in figures 8.1 and 8.2.

Length was originally defined as a fraction of the distance from the North Pole to the Equator, but today it is defined by a measurement of the frequency of the red light from a helium–neon laser and a fixed value of the speed of light (c); so today, the base unit of length is dependent upon the base unit of time (the speed of light is a velocity), whereas, before the 1960s, length and time were completely independent base units. The continued evolution of the SI is required by the continued advance and accelerating pace of advance of science.

Just like any successful natural language, the language of science is forever changing; and just as with the evolution of a natural language, the direction the evolution takes does not always please all those who speak it. However, this evolution is a sign of its health and utility.

Further reading

[1] *Focus on the Revision of the SI* (https://iopscience.iop.org/journal/0026-1394/page/Focus_on_the_Revision_of_the_SI); a source for various publications that contributed to the recent redefinition of several base quantities of the SI, and the creation of the Quantum-SI.

[2] Stock M, Davis R, de Mirandés E and Milton M J T 2019 The revision of the SI—the result of three decades of progress in metrology *Metrologia* **56** 022001

IOP Publishing

Defining and Measuring Nature (Second Edition)
The make of all things
Jeffrey Huw Williams

Chapter 9

20th Century confusions and refinements in measurement

9.1 International politics

For world metrology, the most important event of the 19th Century was the signing of the Metre Convention of 1875. Although the Metre Convention was signed in 1875, which tells us that this event took place under the French Third Republic (4 September 1870–10 July 1940), the terms of the Convention had been created by commissions set up under the previous regime, the Second Empire of Napoléon III. The Second Empire was created by the *coup d'état* of 2 December 1851 and collapsed when the emperor was captured by the Prussians at the Battle of Sedan, 1 December 1870. The elaborate campaign, devised by the commissions of Napoléon III for the metrication of the world, through the instruments which would be created by the Metre Convention was of a scale to rival the military campaigns of his uncle. Slowly but inexorably copies of the prototype standards of the metre and the kilogram were to be manufactured in Paris and dispatched to nations as they signed the Metre Convention. However, this careful plan was perturbed in the late-19th Century and the early-part of the 20th Century by war.

Pan-Germanism (*Pangermanismus* or *Alldeutsche Bewegung*) was a political movement which sought the unity of the German-speaking peoples of Europe. Its modern origins were in the early-19th Century following the Napoleonic Wars and concerned that potent and destructive political creed, nationalism. In the sciences, Pan-Germanism was manifest as a belief that German science was best, which actually was not far from the truth. The development of the chemical and physical sciences in Germany during the 19th Century was far more rapid and advanced than in Britain, France and the USA. French might have been the language of international science and diplomacy in the 18th Century, but the world's most influential scientific language became German in the first-half of the 19th Century and remained German until the mid-1930s, when many of the best scientists left Germany to settle in the English-speaking world.

German science and industry had progressed particularly rapidly in the last part of the 19th Century. The German-speaking peoples of central Europe had been given a new nation by Otto von Bismarck, and the importance of having a single system of weights and measures (the Metric System) was not lost on the industrialists and technocrats who were propelling Germany ever forward. German scientists and engineers were, however, not content with the manner in which the International Prototypes of the Metre and Kilogram were being conserved and copied by the French, who were the guarantors of the Metre Convention of 1875. This lack of confidence by the Germans in the French science was mirrored by a mistrust of the Germans by the French; a mistrust which World War I did nothing to dispel [2 and 3]. The last years of the 19th century saw the rise and fall of many nationalist French politicians who used the populist cry of 'Revenge for 1870–1871' as a means of being elected to office. However, science had to continue and does continue even if nations are at war. We saw earlier how Pierre Méchain, the surveyor of the southern part of the meridian between Dunkirk and Barcelona had been obliged to continue surveying the Pyrenees as the armies of France and Spain battled around the mountains on which he was defining the metre. As Lavoisier put it so succinctly, 'the sciences are not at war.'

Meanwhile, in Great Britain, some Members of Parliament had still not given up on trying to make a metric nation of the UK. Two debates in Parliament in 1907 (under the Liberal administration of Campbell-Bannerman) failed to have the Metric System introduced on a compulsory basis into Britain. One of the arguments voiced against the Metric System during these debates was that 'an agricultural labourer would never ask for 0.568 25 of a litre of beer'. Even though the arguments for defending the *status quo* had reached new heights of drollery, the pro-Metric System Members of Parliament lost the vote.

9.2 Events at the BIPM during the Fall of France, June 1940

This account is taken from a set of unpublished reminiscences (in French) compiled by the BIPM well after the events of 1940 [4].

The BIPM is located in the Parc Saint-Cloud near Paris, yet in 1940 this area was near the site of a large factory manufacturing vehicles (civilian and military). The Renault factory had been a large potential target for the Luftwaffe before France's surrender, and would be a potential target for the Royal Air Force if France were defeated. To prevent damage by aerial bombardment to the metre and kilogram prototypes, the director of the BIPM, Albert Pérard (1880–1960), decided to evacuate the platinum–iridium prototypes from the BIPM. It was clear at the time that as the German army had met no serious opposition on entering France, it was only a matter of days before Paris was captured and occupied. Consequently, a plan was formulated whereby the copies of the International Prototypes, together with the stock of the platinum–iridium alloy, and a 12 kg ingot of gold that the BIPM had acquired from its dotation as a hedge against inflation during the Great

Depression, would all be sent out of Paris in cars belonging to the staff of the BIPM. The destinations were provincial branches of the *Banque de France* in Brittany and the Vendée (Saint-Brieuc and La Roche-sur-Yon, respectively). The only people aware of this plan were the staff of the BIPM and the directors of the banks. It is estimated that at this period, a significant proportion of the population of northern France was on the roads fleeing the advancing German army. So, another couple of un-marked cars (carrying many kilos of platinum–iridium and gold bullion) would go unnoticed in the confusion; but there was an armed guard in each car.

The Metre Convention stipulates (and is reinforced by the decisions of the first few CGPMs) that the International Prototypes of the Kilogram (IPK) and the Metre (IPM) must at all times be kept at the BIPM, which is what happened in 1940. The copies of the prototypes could be hidden away, but the prototypes themselves had to remain on the territory of the BIPM, as stipulated by the Metre Convention.

What the director of the BIPM had to ensure was that the copies of the IPK and of the IPM did not fall into German hands; there was even a suggestion of evacuating the mass standards to the USA, but there was insufficient time. The Germans might take the IPK and the IPM back to Germany, but they would not be able to maintain international metric mass metrology without the copies of the prototypes; it is the copies of the IPK that allows international mass metrology (see section 10.2). Then after the war, assuming Germany lost, which was looking highly unlikely in June of 1940, the copies could be recovered from the anonymous bank vault, and the BIPM could carry on as it had before the war; with France remaining in possession of the metric standards and thus of the dispersal of the Metric System to the world. As it happened, no German soldiers ever ventured onto the international territory that is the BIPM. They always stopped at the gate and asked to see a representative of the BIPM. And in August 1940, with the help of a young German officer, the copies of the International Prototypes were recovered from their bank vaults and repatriated to the BIPM.

The two World Wars and the intervening Great Depression did little to promote the Metric System in Britain, and it was not until the post-World War II economic boom that serious attempts were once again made to bring Britain into the metric family of nations. In July 1959, following a conference of metrologists from the English-speaking nations (the British Empire, Commonwealth and the USA), the participants agreed to unify their standards of length and mass and to define them purely in terms of metric units. Before this conference, the UK inch had measured 25.399 8 mm, while the USA inch was ever so slightly longer at 25.400 05 mm, but the difference in units of volume between the UK and the USA were even starker. However, after 1959 the English-speaking nations agreed to adopt the inch as defined by the new International Standards Organization (ISO) of Geneva; that is, an inch equal to 25.4 mm. The National Bureau of Standards (today called the National Institute of Standards and Technology) in the USA reported that 'As a result of many years of preliminary discussion, the directors of the national standards laboratories of Australia, Canada, New Zealand, South Africa, the United Kingdom, and the United States [have] entered into agreement, effective July 1, 1959, whereby uniformity was established for use in the scientific and technical fields. The equivalents one yard to be equal to 0.9144 metre (whence one inch is equal to 25.4 millimetres)...'.

A new Weights and Measures Act was passed by the British Parliament in 1963. This Act defined the yard and the pound in terms of the metre and the kilogram, respectively; the yard was now equal to 0.9144 m and the pound was equal to 0.453 592 37 kg. From this date, the old Imperial quantities have been legally dependent on the SI standards of the metre and kilogram. Britain was moving towards... 'inching' towards the Metric System.

The Metric System was not being adopted by the English-speaking world, as this would have been a political statement, and in the late-1950s when the USA led the world in everything, the Americans could not be seen to be adopted a system of weights and measures which had been invented by another nation, and was maintained by that other nation. However, the non-metric nations were prepared to attempt a standardization of their own weights and measures and bring them into closer alignment with the Metric System. This would be an essential first step towards a full and complete future metrication and to improving trade between all nations. The position of Britain in the world was also changing; not only was its empire evaporating, we were also moving our systems of weights and measures so as to facilitate a future change to the Metric System. Then in 1961, Greenwich Mean Time which had been the basis of the world's time coordination was renamed Coordinated universal Time (UTC) by international agreement.

9.3 Two peoples separated by a common system of weights and measures

One of the emptiest arguments used by anti-metric politicians in the UK and the USA during debates about a full adoption of the Metric System in the mid-20th Century was that *we* cannot adopt the Metric System because *they* have not adopted it; the 'we' and the 'they' are interchangeable for the British or US governments. Sadly, these individuals did not appreciate that although there are quantities and units in the UK and the USA which share the same name, they do not necessarily have the same value.

British weights and measures, or more precisely English units as they derive from the units of measurement used in England up to 1824 (the other countries that occupy the British Isles had their own customary units[1]) evolved as a combination of Anglo-Saxon and Roman weights and measures. The first national system of weights and measures in Great Britain was created by the redefinition undertaken by Parliament in 1824, which repealed nearly all previous legislation related to weights and measures.

---

[1] Welsh units of measurement were those used in Wales between the Sub-Roman period (prior to which the Britons used Roman units) and the 13th Century Edwardian conquest (after which English units were imposed). In the Venedotian Code used in Gwynedd, the units of length were said to have been codified by Dyfnwal Moelmud but retained unchanged by Hywel Dda (c.880–948). The code provided for a variety of means of defining units, and for deriving them from grains of barley (see chapter 1).

While it is true that the customary units of the UK and the USA have a shared heritage, there are significant differences. The US customary measurement system is based on the English customary measurement system of the late-18th Century, while the British imperial system of weights and measures was created in 1824; that is, well after American independence.

This difference is seen most starkly in measurements of volume. The American colonists adopted the English wine gallon of 231 cubic-inches as a standard. At that time, the English also used this wine gallon for measuring volumes of other liquids, but they also used the ale gallon of 282 cubic-inches for measurement of volume. The imperial gallon of 1824 was about 277.42 cubic-inches, which made it much closer to the volume of the English ale gallon than to the volume of the English wine gallon, which was by 1824 the standard volume for liquids in the USA. This imperial gallon, which is approximately 4.55 l, is 20% larger than the US liquid gallon (3.785 l). This difference carries through to other units of liquid volume, some of which are still used on both sides of the Atlantic, and extends into quantities used to define mass that are related to volume.

Although early signatories of the Metre Convention, neither Great Britain nor the USA had adopted the Metric System by 1900. Two World Wars and the intervening Great Depression did little to promote the Metric System in Britain or the USA[2], and it was not until the post-World War II economic boom that serious attempts were once more made to bring the English-speaking world into the metric family of nations.

In 1978, the SI became the only legally recognized system of units in the European Economic Community, the European Union of today. The necessary legislation for cut-off dates for the use of the new units, drafted by the UK Board of Trade to give the UK time to organize the change from British units to SI units, was to be agreed by legislation arising from a nation-wide consultation. However, this initiative was in the hands of Secretary of State for Trade who chose not to test the opinion of Parliament by a debate of the proposed legislation, which he withdrew. The General Election of 1979 brought about a change of government, and given the antipathy of the new Prime Minister, Margret Thatcher, and her supporters for anything to do with the European Union, the metrication of the UK was no longer a priority issue.

However, the advance of science does not wait upon the occupant of Downing Street. In 1980, the wavelength of an iodine-stabilized helium–neon laser was accepted by the CGPM as the wavelength which would define the metre, or length standard of the SI. A laser source of the light used to realize the metre offered significant advantages over the krypton atomic lamp, which had been adopted as a standard in 1960. Such a stabilized red helium–neon laser had a wavelength uncertainty of 1 part in 10^{10}.

In 1983, the 17th CGPM further refined the definition of the metre 'as the length of the path travelled by light in a vacuum during a time interval of 1/299 792 458 of a second' (17th CGPM, Resolution 1). This new definition assumed that the speed of

---

[2] Details of when nations signed the Metre Convention, and then finally adopted the Metric System may be found on the website of the BIPM; www.bipm.org.

light, c, was now fixed at $c = 299\ 792\ 458$ m s^{-1} exactly. As always when the CGPM changed the definition of one of the base units of the SI, the goal was not only to improve the precision of the definition of that unit, but also to change its actual value (that is, the length of the metre) by as little as possible.

The new definition of the metre relied upon a fixed speed of light, which defined the distance light travels in a particular period of time; so many metres per second. Then, by defining a period of time, one could realize the distance that the light had travelled in that period of time, and one could thus define the unit of length by means of a unit of time. The unit of time, the second, could be determined to an extraordinary level of precision, 1 part in 10^{14}, by using a caesium atomic clock.

While this revolution in the nature of the very small and the very precise was unfolding, further moves were made to bring the UK into the Metric System. The UK had been a member of the European Union since 1973, and in 1989, the European Union issued a binding Directive on Weights and Measures to Member States. This Directive declared that the SI would be the official measurement system of all Member States of the European Union, and hence of the UK, and that, as far as the UK was concerned, this was now a matter of international law, and not of individual choice. The main effect of this Directive was to mandate the use of metric units for all pre-packed goods by the end of 1994, and for all bulk goods to be priced in terms of metric units by the end of 1999. The Directive placed pressure on the UK to finally fully adopt the Metric System. (Figure 9.1 shows a 1975 medallion manufactured to celebrate the centenary of the Metre Convention of 1875. It shows the recent redefinition of the metre in terms of the wavelength of light.)

Figure 9.1. Medal commemorating the centenary of the Metre Convention and the BIPM, manufactured by R Corbin, *Monnaie de Paris*. This face of the medal represents the seven base units of the SI (metre, kilogram, second, ampere, kelvin, mole and the candela), and how the metre is defined in terms of the wavelength of light (in 1975 this was via the red light from a Krypton discharge lamp) rather than by an artefact. (This image is reproduced with the permission of the BIPM, which retains full international copyright.)

In 1995, Parliament passed legislation to incorporate the 1989 EU Directive into UK law. This new legislation stated that 'for economic, public health, public safety, and administrative purposes', only SI units of measurement would have any legal validity in the UK. This meant that all packaged goods were required to be labelled in metric units. However, it was agreed with the EU Commission that until 31 December 1999: pounds and ounces could be retained for the weighing of goods sold in bulk in markets; pints and fluid ounces be used for alcoholic beverages such as beer and cider, water, lemonade and fruit juices in returnable containers; therms for gas supply; and fathoms and nautical miles for marine navigation.

In addition, it was agreed that some measures would not be subject to any time limit, and so would never be converted to metric units. These quantities are: statute miles, yards, feet and inches for road traffic signs and related measurements of speed and distance; pints for dispensing draught beer and cider, and for milk in returnable containers; acres for land registration purposes and surveyors' measurements; and troy ounces for transactions in gold and other precious metals.

Interestingly, while the UK is presently pre-occupied with Brexit, we have seen here how the CGPM has moved on and redefined the base quantities of the SI to create a Quantum-SI. This new formulation of the SI has already been adopted by the EU; by the Directive (EU) 2019/1258 of 23 July 2019 (amending, for the purpose of its adaptation to technical progress, the Annex to Council Directive 80/181/EEC as regards the definitions of SI base units). Thus, as far as the scientific world is concerned, the new Quantum-SI is the only formulation of the SI to be used. And this is now the case with the wider society of the EU. For the various technical bodies of the EU, the kilogram is no longer defined by an object in Sèvres, but by the value of the Planck constant. However, until such time as the UK government decides how closely aligned with the EU they wish to be, and which EU legislation they will introduce into British law, the kilogram in the UK will continue to be defined by the IPK in Sèvres.

Thus, we have the curiously muddled system of units seen today in Britain. A strange mixture of units, but a mixture which strongly resembles the situation in France after Napoléon Bonaparte abolished the Metric System, for domestic political reasons, in France in 1812.

Further reading

[1] For a history of how the Metric System has been accepted around the world, see *Le Mètre du Monde*, Denis Guedj, mai 2000; Editions du Seuil.
[2] *Le Système Métrique; est-il en Péril?* J Rene-Benoit et Ch-Ed Guillaume, *La Révolution Française* 1916 (Juillet–Aout) pp 3–26; *Le système métrique; est-il en péril?* Ch-Ed. Guillaume, *L'Astronomie* (1916) **30** 242–9.
[3] The two publications in [2], defending the status quo of the Metric System in 1916 were published in response to this article: *Pangermanisme et système métrique*, G Pariset, '*La Révolution Française*' 1916 (Janvier–Février) pp 5–34.
[4] *Je me souviens*, in *Nous nous Souvenons* (unpublished personal reminiscences of the career of M Albert Bonhoure, who worked at the BIPM from 1912–1963; Library of the BIPM).

IOP Publishing

Defining and Measuring Nature (Second Edition)
The make of all things
Jeffrey Huw Williams

Chapter 10

The birth of the Quantum-SI

To comply with the terms of the Metre Convention of 1875, the first *Conférence générale des poids et mesures* (CGPM) of 1889 formally approved the manufacture of 40 prototype metres and 40 prototype kilograms by the British firm, Johnson Matthey. These were to be the metrological standards mandated by the Metre Convention; to be used for the metrication of the world. Once these artefacts had been sent to each member state of the Convention, that nation would be in a position to set up a national system of metrology based on the metre and the kilogram. Each of these new national systems of metrology would then be traceable to the International Prototypes of the Kilogram and the Metre kept in Sèvres.

To create this pyramid of metrological traceability, one of each of the newly manufactured standards was nominated by lot as the International Prototypes of the Kilogram and the International Prototype of the Metre. The CGPM retained some of the kilogram artefact and of the metre artefact as working copies, or *témoins*, and the remainder of the prototypes, manufactured by Johnson Matthey were distributed to the member states of the Metre Convention. At regular intervals the various national prototypes were compared with, and recalibrated against the international prototypes, which were kept at the *Bureau international des poids et mesures*, Sèvres.

In 1921, the Metre Convention was revised, and the mandate of the CGPM was extended to provide standards for all units of measurement, not just for mass and length. Subsequently, the CGPM took on responsibility for providing the highest metrological standards of: electrical current (1946), luminosity (1946), temperature (1948), time (1956), and molar mass (1971). Then the 9th CGPM of 1948, instructed the *Comité international des poids et mesures* (CIPM) 'to make recommendations for a single practical system of units of measurement, suitable for adoption by all countries adhering to the Metre Convention.' The recommendations based on this mandate were presented to the 11th CGPM of 1960, where they were formally accepted, and given the name '*Système International d'Unités*' with its abbreviation SI.

doi:10.1088/978-0-7503-3143-2ch10
10-1

10.1 The need for change

These first changes to the terms of the Metre Convention provided a precedent, and basis for changing the underlying principles behind the definitions of the base units of the SI. Subsequently, there have been several changes to the definitions of the base units of the SI. The 11th CGPM of 1960 defined the SI metre in terms of the wavelength of red krypton-86 radiation from an atomic discharge lamp, thereby replacing the pre-SI metre bar, or artefact; and the 13th CGPM of 1967 replaced the original definition of the second, which was based on Earth's average rotation, with a definition based on the frequency of a particular transition between two hyperfine levels of the ground state of the caesium-133 atom. The 17th CGPM of 1983, replaced the 1960 definition of the metre, with one based on the second by giving an exact definition of the speed of light, c, in units of metres per second.

At this time, the SI was a coherent system constructed around seven base units. Various combinations and powers of these base units were used to construct all other units; indeed, would be used to explain phenomena as yet unobserved. With the 2019 redefinition of the base units of the SI, the SI is now constructed around a defining set of constants of Nature, allowing all units to be constructed directly from these constants. The designation of base units is retained, but they are no longer essential to define the SI quantities. The SI has moved from being defined by macroscopic quantities (for example, a glass water triple-point-cell containing many cubic centimetres of water) to being defined by constants of Nature.

10.2 The problem that was the kilogram

The recently abolished hierarchy of measurement standards for mass closely resembled a religious dogma. At the highest point was the omnipotent object which was, as far as the world is concerned, indivisible and in which one had to believe and have faith, but without ever touching, or even seeing it. The perfect mass of this object, precisely one kilogram, was fixed by international treaty and not by mere experiment.

This near-sacred object was (and still is) the International Prototype of the Kilogram (IPK). This memento from the early-days of the Metric System was one of science's most valuable, but also one of its most derided objects. It is kept in a secure, subterranean vault near Paris surrounded by six identical copies or *témoins* (see figure 10.1). These platinum–iridium prototypes were used every half century or so to monitor how the stability of these mass artefacts was being maintained and to provide calibration certificates to the rest of the metric world so that we may all see that 'the kilogram' was in its vault, and that 'all was right with the metric world'.

The good news of the continued stability and perfection of 'the kilogram' then passes down from the closely guarded objects in their French vault to their copies in the various member states of the Metre Convention and then down the various national pyramids of scientific endeavour and precision measurement to, for example, the humble weighing-scale in your bathroom. In this way, you may be reassured that you are not fooling yourself after a holiday of over-indulgence; as your weighing-scale has certificates of calibration linking it to the perfect essence of a

Figure 10.1. The author standing in the vault at the BIPM, in front of the open strong-box which contains the International Prototype of the Kilogram (middle shelf) and its six copies or *témoins* (upper and lower shelves). These seven mass standards were the 'Crown Jewels' of international mass metrology; that is, they were the artefacts whose fluctuating masses defined the world's system of mass metrology.

kilogram locked away at the BIPM in Sèvres. This is essentially the way any artefact-based system of measurement works. There are local standards traceable to national standards, and these national standards are traceable to international standards. And finally, the highest standards are traceable to the standard of that unit; the Platonic essence of that unit.

Many years ago, one of the things I learnt early on when telling friends that I was working at the BIPM was that they had all heard that the kilogram was losing...or was it gaining weight uncontrollably. My friends had read dramatic reports in the popular press, which had commented that the kilogram was not all it should be, and was actually probably less than it should be. Such reports appeared quite regularly; usually to correspond with one of the meetings of the General Conference on Weights and Measures.

Of course, such statements beg the question, which is never addressed; getting lighter compared to what? The inference is that the kilogram is not a sound unit. That it is getting lighter compared to a more stable mass. But how can this be determined? If there were a more stable mass, it would be the International Prototype of the Kilogram, and the present international prototype would become just another mass standard. Given the artefact basis of mass metrology at that time, some object had to occupy that solitary, stable position at the apex of the pyramid of the traceability of mass measurements. The possible instability in the physical characteristics of a metrological artefact is the essential problem with all artefact-based systems of weights and measures; which is why they are re-calibrated every few years.

During my time at the BIPM, I saw the IPK on several occasions; to be precise, I walked past the open door of the strongbox and looked in and saw the IPK under its three protective glass bell jars (see figure 10.1). Very few people get closer than that.

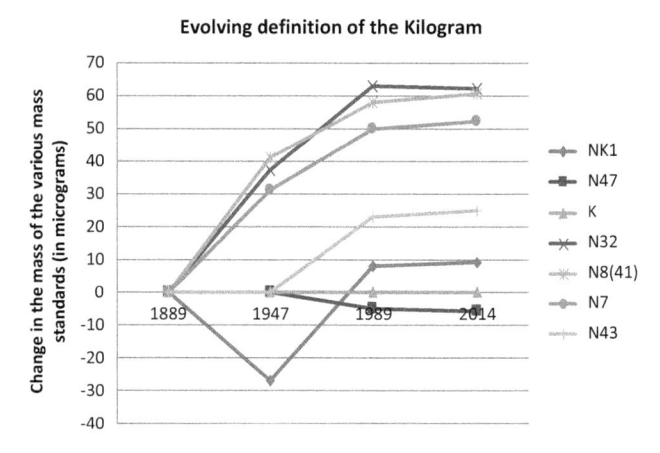

Figure 10.2. A graph of the relative change in mass of selected kilogram prototypes. Some of the data is taken from Girard G 1994 *The Third Periodic Verification of National Prototypes of the Kilogram (1988–1992) Metrologia* **31** 317–33, and the most recent values (2014) are taken from the BIPM website (https://www.bipm. org/en/bipm/mass/ipk/). The International Prototype Kilogram (IPK) kept in Sèvres, France, and the other national prototype kilogram standards have been found to vary in mass over the years. Since the IPK defines the kilogram, the only way to detect changes in mass is to compare the mass of the prototypes with the IPK (the mass of which is fixed by definition), which is done every 40–50 years. This graph shows the 'mass drift' of six numbered prototypes compared to the IPK.

It is a bit like going to see the Crown Jewels in the Tower of London. One may look, one may be awed and overwhelmed, but one may not linger—let alone touch. What most impressed me about the IPK is its apparent tangibility, its solid, sculpted form of a shiny alloy of platinum and iridium. As you briefly looked into the strongbox, you realized that this was not just any mass standard; this was the quintessential kilogram. Manufactured in 1886, and defined by a diplomatic treaty to be the most stable mass in the world.

The IPK led a sheltered life. It is kept inside a temperature- and humidity-controlled vault in a secure room within the *Parc de Saint-Cloud* enclave of the BIPM. It has been on this site during two world wars when France was attacked and occupied by another member state of the Metre Convention. Thus protected, it reigned supreme over the world's measurements of mass. Every hill of beans, every human, every milligram of medicine and every little bag of recreational drug; in short, the great globe itself and even the smallest of sub-atomic particles that can be weighed had to be compared against the mass of that small, glittering, dense object.

10.2.1 The 'smoking gun'

Since it was placed in service in 1889, the IPK has been used during three measurement campaigns or 'periodic verifications of national prototypes of the kilogram'. The results of the most recent such verification were published in 2014. The temporal stability of the masses of the official copies of the IPK with respect to the IPK itself is shown in a well-known graph given below (figure 10.2) and which may be found in detail on the BIPM's website (www.bipm.org/en/scientific/mass/verifications.html) [1]. The figure shows the change in the calibrated mass with respect to time for the copies of the IPK compared with their initial calibration in 1889. Over a period of more than a century, the masses of the official copies are seen to be increasing, with respect to the mass of the IPK, which by definition cannot change its mass and so appears as a horizontal line in the figure labelled K.

By definition, the mass of the IPK cannot vary, but given that the masses of its copies are fluctuating, then the mass of the IPK must also be changing, because the IPK and its copies are all stored, and have been stored in the same manner; they are also made of the same material and were all made in the same manner. And given that the masses of the copies are mostly increasing, then as they are weighted against the IPK, the mass of the IPK must also be decreasing. This is the conclusion so beloved by science journalists seeking a 'story'.

Unfortunately, for the scientists who rely on the continued stability of the base unit of mass for precise measurements, this inconstant metric mass is a nuisance [1]. Our ability to precisely measure an electric current or a quantity of gas flowing through a pipeline is dependent upon the precision with which the unit of mass is known, and any instability in the precision with which we are able to define the base unit of mass perturbs such calculations; calculations which are worth hundreds of billions of euros every year.

Since their manufacture in the late-1890s, the masses of the kilogram artefacts were seen to be drifting. Drifts of up to 2×10^{-8} kilogram per year in the national prototype kilograms, relative to the International Prototype of the Kilogram (IPK) had been detected. There is no way of determining whether the national prototypes were gaining mass, or whether the IPK was losing mass. Research has since identified mercury vapour absorption or carbonaceous contamination (city pollution) as possible major causes of this drift. At the 21st CGPM of 1999, national laboratories were urged to investigate ways of severing the connection between the SI unit of mass, the kilogram and a specific artefact [1].

The IPK is today, by virtue of the recent redefinitions of the base units of the SI, a museum piece. But prior to 20 May 2019, the IPK was the only object to have its mass determined exactly. It possessed a mass of precisely 1.0 kg, because that mass was defined by international law (related to the Metre Convention of 1875) as being precisely 1.0 kg, and not by mere experiment. By comparison, all the other mass standards would have had a unique, measurable mass, different from 1 kg. Yet, for all the sparseness of the data presented in figure 10.2, this image carried a huge responsibility. The IPK will, of course, always have had a mass different from 1.0 kg, but it could not be measured. The values of mass along the vertical axis in figure 10.2 give the range of values within which one would be likely to discover the true mass of the IPK. Today, the mass of the IPK can be measured (and probably will be measured to show the scientific world why this redefinition was needed), as the kilogram has been redefined by a non-artefact-based definition, and so the IPK has become just a lump of platinum–iridium alloy.

So how much was a kilogram? Well, as it turns out nobody could say for sure; at least, not in a way that will not change ever so slightly over time. And that was of no use to scientists who are interested in determining the mass of small objects; for example, the recently discovered Higgs Boson; and then there is the problem of the mass of the neutrino—is it zero, or finite? Before 20 May 2019, given the artefact basis of mass metrology, some object had to occupy that solitary position at the apex of the pyramid of mass measurements. The instability in the physical characteristics of a metrological artefact is the essential problem with all artefact-based systems of weights and measures, which is why they require regular recalibration. The creation of the quantum-SI has removed the need for such regular recalibrations.

The SI is the pivot from which hang all measurements, no matter what the area of investigation or the location of the measurement. Whether it is the accuracy of your bathroom scales, the amount of electricity you have consumed in the last month, or the reliability of a petrol pump, there is an unbroken chain of calibration certificates that leads back to realizations of the seven base units of the SI. Science has moved a long way since the IPK became the basis of mass metrology, and this object increasingly came to be seen as anachronistic; as can be seen in the carefully measured data in figure 10.2. The surest way of stabilizing mass metrology, and to remove our dependence on the drifting values of the masses of a collection or artefacts was to do away with the IPK altogether. Greater precision would also be brought to the SI, if the unit of mass were defined by a constant of Nature rather than an object; no matter how carefully conserved. The choice of the physics community was to redefine the unit of mass in terms of Planck's constant.

10.3 The background to the redefinition

It can thus be seen that by the end of the last century, the metrology community was seeking an alternative basis for the definition of mass. But what was the route that lead from the IPK to the definition of mass in terms of the Planck constant?

It was quickly decided that the redefinition of the kilogram was not the only problem facing the SI, and that a more substantial reform of the entire SI, might be required to ensure its continued scientific relevance. As laid down by the Metre Convention, which created international institutions specifically to guarantee the coherence of international systems of measurement, and the mechanism of how these systems should evolve, it is through the *Conférence générale des poids et mesures* (CGPM) that the first call about a future SI was made to the wider metrological community. This call from the CGPM led to several alternative approaches to redefining the kilogram, based on fundamental physical constants. Among others, the Avogadro project and the development of the Kibble balance (originally known as a watt balance, before 2016) promised methods of indirectly measuring mass with high precision. These projects also provided tools that would enable alternative means of redefining the kilogram.

A report published in 2007 by the Consultative Committee for Thermometry (CCT) of the BIPM, to the CIPM noted that the then current definition of temperature had proved to be unsatisfactory for temperatures below 20 K, and for temperatures above 1300 K. The committee took the view that the Boltzmann constant provided a better basis for temperature measurement, than did the triple-point of water because it overcame these specific difficulties. Finally, at the 23rd CGPM in 2007, the CIPM was mandated to investigate the use of the constants of Nature as the basis for all the base units of the SI. The following year, this decision was endorsed by the International Union of Pure and Applied Physics (IUPAP).

To begin the process of creating a new-SI, a Quantum-SI in 2010, the Consultative Committee of Units (CCU) began a revision of the 8th edition of the *SI Brochure*. The CCU was keen to press on as quickly as possible with remaking the SI. However, the CIPM meeting of October 2010 found 'the conditions set by the General Conference at its 23rd meeting [for the redefinition of the base units] have not yet been fully met. For this reason the CIPM does not propose a revision of the SI at the present time'. The CIPM, however, presented a Resolution for consideration at the 24th CGPM (October 2011) to agree to the new definitions 'in principle', but not to implement them until the details had been finalised. Essentially, the CIPM was keen that there should be transparency in this process of remaking the SI, and pressed for the creation and publication of a 'road map', which would show the wider-world how a small group of international metrologists were proposing to redefine the units upon which humanity was dependent. In this way, the difficulties of communicating the changes to the wider society would be lessened; the change would not seem to be a *fait accompli*.

This Resolution from the CIPM was accepted by the 24th CGPM. In addition, this CGPM moved the date of the 25th CGPM forward from 2015 to 2014. However, at the 25th CGPM (November 2014), it was again found that 'despite [progress in the necessary requirements] the data do not yet appear to be sufficiently robust for the CGPM to adopt the revised SI at its 25th meeting'. Thus further postponing the redefinition of the SI to the next CGPM in 2018. Measurements accurate enough to meet the conditions were finally available in 2017, and the redefinition was adopted at the 26th CGPM in November 2018; with the changes taking effect on 20 May 2019, World Metrology Day—the anniversary of the Metre Convention.

But what of the trigger, which instigated the final push that led to these recent redefinitions? What was it that started the various international bodies, created by the Metre Convention of 1875, to start the process of creating a Quantum-SI? The firing of the starting pistol can probably be traced to a paper of 2005 [2], *Redefinition of the kilogram: a decision whose time has come*. The authors point out how the watt balance, as was (today, the Kibble balance[1]) can be used to realize the unit of mass, the kilogram, by fixing the numerical value of the Planck constant. This article in *Metrologia* explains the Kibble balance, its role in the redefinition of the unit of mass, and attempted to define a future-SI. Indeed, the five authors were in the vanguard of those seeking a recreation of the SI. However, the wider international metrology community, who were no less in favour of creating a Quantum-SI were keen to demonstrate the breadth of the consultations that had led to the coming

---

[1] In 2017, the watt balance was renamed to the Kibble balance to honour the inventor, Bryan Kibble of the NPL, Teddington, who had died in 2016.

changes. The redefinition of the SI, to create a Quantum-SI represents an enormous achievement of a great many scientists from around the world; all working to improve the art of measurement.

The Kibble balance, see section 11.3.1 for full details, is one of two possibilities that can be used to realize the unit of mass at the kilogram level with relative uncertainties of approximately 10^{-8}. The other possibility is the x-ray crystal density (XRCD), which is discussed in section 13.5.

The idea of the Kibble balance was published in 1976 [3]. The original Kibble balance was used to realize the ampere; the base unit for amount of electricity of the SI. However, the use of the Kibble balance evolved with the wide use of quantum mechanical phenomena in electrical metrology; in particular, with the introduction of the quantum Hall effect, discovered in 1980 by Klaus von Klitzing, and the quantum effect predicted earlier by Brian Josephson. Via these quantum phenomena, the Kibble balance morphed from a device capable of realizing the SI unit of current, to an experiment that could measure the Planck constant, h, with high precision. It became an instrument that could and will be used to realize the SI unit of mass. This was the proposal put forward by the five authors in 2005 [2][2].

The Kibble balance is a complex device that connects classical mechanics to quantum mechanics. In detail, the Kibble balance achieves this bridging of the two main areas of physics by dividing the measurement and analysis into two parts; firstly, one relates mechanical quantities to electrical quantities, and then the electrical quantities to the quantum mechanical phenomena.

Further reading

[1] Stock M, Barat P, Davis R S, Picard A and Milton M J T 2015 Calibration campaign against the international prototype of the kilogram in anticipation of the redefinition of the kilogram part I: comparison of the international prototype with its official copies *Metrologia* **52** 310–6
Davis R S, Barat P and Stock M 2016 A brief history of the unit of mass: continuity of successive definitions of the kilogram *Metrologia* **53** A12–8

[2] Mills I M, Mohr P J, Quinn T J, Taylor B N and Williams E R 2005 Redefinition of the kilogram: a decision whose time has come *Metrologia* **42** 71–80

[3] Kibble B P 1976 A measurement of the gyromagnetic ratio of the proton by the strong field method *Atomic Masses and Fundamental Constants 5* ed J H Sanders and A H Wapstra (Boston, MA: Springer) 545–51

---

[2] The author was the editor of *Metrologia* when the manuscript by Mills *et al.* was submitted. Some years after the paper had been published, one of the referees who had been asked for an opinion on the submission told the author, while smiling that given all the subsequent furore, perhaps it would have been better to insist that a question mark be inserted at the end of the title.

Chapter 11

The base units of the Système International des Unites (I)

So far in this volume, we have seen something of the history of metrology and, in particular, a history of the Metric System, known today as the International System of Units (SI). We will now consider the present status of the seven base units, which are combined, as shown earlier, to generate all the derived units required to measure and quantify Nature. In particular, we will look at the recent redefinitions of the base units of the SI in terms of constants of Nature. Given that the recent redefinitions are so very new; we will in the coming chapters consider both the new definition and the old definition for each of the seven base units. The two sets of definitions are summarised in table 11.1.

11.1 The base unit of length is the metre (m)

Originally, the metre was defined by an artefact, or measurement standard. The definition of the metre from the 1st CGPM of 1889 was based on such an artefact, the International Prototype of the Metre made of an alloy of platinum and iridium, but it was replaced by the 11th CGPM of 1960 with a definition based on the wavelength of the red-light emitted by excited krypton atoms in an atomic discharge lamp.

Although the stability of the original artefact prototype was verified by comparison with three companion bars, there was from an early-date a move to use the speed of light to define distance. There were nine measurements of the metre using the red-light emitted by excited atoms of cadmium in discharge lamps between 1892 and 1942. The first of these measurements was carried out by the American physicist Albert Abraham Michelson (1852–1931) using the interferometer which he designed, and for which he received the Nobel Prize in physics in 1907.

The Michelson interferometer is the simplest optical device for making precision measurements of optical frequencies (that is, the speed of light divided by the optical

Table 11.1. Definitions of the base units of the SI before 20 May 2019 (the Old-SI), and subsequent to that date (the Quantum-SI). See text for further details, and figures 8.2 and 8.1, respectively, for the interconnectivity of the base units.

Quantity (base unit)	Definition pre-May 2019	Present definition
		The International System of Units, the SI, is the system of units in which:
Metre The base unit of length is the metre (m)	The metre is the length of the path travelled by light in vacuum during a time interval of 1/299 792 458 of a second	The speed of light in vacuum c is 299 792 458 m s^{-1}.
Kilogram The base unit of mass is the kilogram (kg)	The kilogram is equal to the mass of the International Prototype of the Kilogram (see figure 7.4)	The Planck constant h is 6.626 070 15 × 10^{-34} J s.
Second The base unit of time is the second (s)	The second is the duration of 9 192 631 770 periods of the radiation corresponding to the transition between the two hyperfine levels of the ground state of the caesium-133 atom. (This definition refers to a caesium atom at rest at a temperature of 0 K.)	The unperturbed ground state hyperfine transition frequency of the caesium-133 atom $\Delta\nu_{Cs}$ is 9 192 631 770 Hz
Ampere The base unit of electric current is the ampere (A)	The ampere is that constant current which, if maintained in two straight parallel conductors of infinite length, of negligible circular cross-section, and placed 1 metre apart in vacuum, would produce between these conductors a force equal to 2 × 10^{-7} newton per metre of length	The elementary charge e is 1.602 176 634 × 10^{-19} C
Kelvin The base unit of thermodynamic temperature is the kelvin (K)	The kelvin, the unit of thermodynamic temperature, is the fraction 1/273.16 of the thermodynamic temperature of the triple-point of water. (This definition refers to water of the isotopic composition: 0.000 155 76 mole of H-2 per mole of	The Boltzmann constant k_B is 1.380 649 × 10^{-23} J K^{-1}

	H-1, 0.000 379 9 mole of O-17 per mole of O-16, and 0.002 005 2 mole of O-18 per mole of O-16.)	
Candela The base unit of light intensity is the candela (cd)	The candela is the luminous intensity, in a given direction, of a source that emits monochromatic radiation of frequency 540×10^{12} hertz and that has a radiant intensity in that direction of 1/683 watt per steradian	The luminous efficacy of monochromatic radiation of frequency 540×10^{12} Hz, K_{cd}, is 683 lm W^{-1}
Mole The base unit of amount of substance is the mole (mol)	The mole is the amount of substance of a system which contains as many elementary entities as there are atoms in 0.012 kg of carbon-12	The Avogadro constant N_A is $6.022\ 140\ 76 \times 10^{23}$ mol^{-1}

wavelength). Michelson, together with Edward Morley, used this interferometer for the well-known Michelson–Morley experiment (1887), which demonstrated the constancy of the speed of light across multiple inertial frames; that is, with light travelling in different directions relative to the rotation of the Earth. This experiment famously removed the conceptual need for a luminiferous aether to provide a rest frame for light.

The 11th CGPM of 1960, defined the metre as the length equal to 1 650 763.73 wavelengths of the red-light emitted from excited Krypton atoms. Subsequently, improvements in light sources; particularly, after the discovery of the laser in the 1960s generated new optical standards based on well-defined wavelengths of light.

In 1975, the CGPM recommended a value for the speed of light as a result of measurements of the wavelength and the frequency of the light emitted by stable lasers. And in 1983, the CGPM redefined the metre as the length of the path travelled by light in a vacuum during a specific fraction of a second. The CGPM then asked the CIPM to draw up a set of instructions for the practical realization of the new definition of the metre; and the wavelengths, frequencies and associated uncertainties were subsequently specified by the CIPM in the instructions for the practical realization of the definition of the metre. This list of recommended radiations was first published in 1983 in the *mise en pratique* of the definition of the metre (see www.bipm.org/en/publications/mep.html), see figure 11.1 for details.

This move from an artefact-based metre to an optical metre came about to improve the precision with which the metre could be realized; the realization being achieved using an interferometer with a travelling microscope to measure the optical

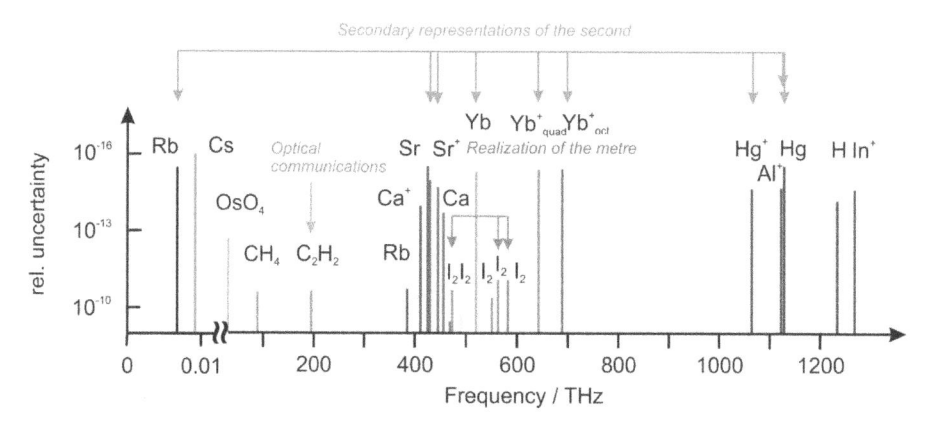

Figure 11.1. Frequencies of radiations used in the realization of the metre, the second (Cs), and (presently) in secondary representations of the second. This graph is reproduced, with kind permission from Riehle *et al* 2018 *Metrologia* **55** 188 (figure 11.3) copyright 2018 BIPM & IOP Publishing Ltd, CC BY 3.0.

path difference as the fringes produced in the interferometer were counted. These various changes of definition reflect an increasing need from the scientific community to have the universal measure known to ever-increasing levels of precision. And they served as a basis for the recent reformulation of the SI as the Quantum-SI.

As pointed out earlier, the present definition of the metre does not rely on a measurement of distance per se, but upon the distance travelled by the red-light emitted by a He–Ne laser. To realize the metre in this manner, it is necessary to fix the speed of light, which is consequently defined, or fixed at the same time that the metre was redefined, to be exactly 299 792 458 m s^{-1}, thus $c = 299\ 792\ 458$ m s^{-1}. Experimental procedures which were previously interpreted as measurements of the speed of light have now become calibrations of length, and a fundamental, unchanging property or constant of Nature (the speed of light) has become the basis for the definition of a macroscopic observable, the metre (see table 11.1).

Alternatively, we can ask, what is the distance travelled by light in 1/299 792 458 s? Well, that is precisely one metre. This definition means that the speed of light has been fixed. Albert Einstein said that the speed of light was the fastest possible velocity, and metrologists, in order, to better define the metre have fixed the speed of light at the value given above. Consequently, one cannot use the relationship between speed of light and wavelength (which is a distance) to measure the speed of light, because the two quantities are no longer independent. If future, scientists wishing to determine a new, more precise value of the speed of light must find some new phenomenon from which to make the measurement and the determination, and this new value of the speed of light will then overturn the accepted definition of the metre—they must seek new paradigms.

In this manner, the metre, the SI base unit of length may be realized in any competent national laboratory, and is not tied to an artefact keep in the French national archives, nor is it tied to the dimensions of the Earth as it was in the late-1790s, which are not perfectly constant. The Earth is constantly changing on readily measurable scales, sometimes over very short periods. The 2011 Tohoku earthquake, for example, was sufficiently powerful (magnitude 9.0 M_w) to have moved Honshu, the main island of Japan, by 2.4 m in an easterly direction, and to have shifted the Earth on its axis by between 10 cm and 25 cm.

11.2 The base unit of mass is the kilogram (kg)

As seen in table 11.1, there was absolutely no possibility for confusion about the definition of the kilogram in the old SI. This definition stated unequivocally that an object kept in a safe in a vault at the *Bureau international des poids et mesures* (BIPM) near Paris weighed exactly one kilogram. No more, no less; one kilogram precisely. This was the last example of an artefact-based definition in the SI, and the last echo of the original Metric System of April 1795.

The International Prototype of the Kilogram (IPK) is an artefact, or metrological standard, made of an alloy of platinum and iridium, which has been kept at the BIPM under the conditions specified by the 1st CGPM of 1889, when it was sanctioned as the international prototype and to be the unit of mass of the Metric System. The IPK is a cylinder of a dense alloy with equal height and diameter of 3.9 cm, and with slightly rounded edges. For a cylinder, these dimensions represent the smallest surface area-to-volume ratio so as to minimize wear. The mass of the IPK is certainly about one kilogram, but one could not actually measure its mass as there is nothing to weight it against—it being 'the international standard', and so it was the object with the most precisely defined mass (defined by a Resolution of the GCPM, which is backed by the authority of the Metre Convention of 1875 and associated man-made laws). Thus, there was no experimental realization of this base unit; the artefact known as the International Prototype of the Kilogram was the physical manifestation of the base unit of mass.

Before continuing, let us consider an often-confused difference. Weight and mass are fundamentally different quantities; mass is an intrinsic property of matter, whereas weight is a force that results from the action of gravity on mass.

The weight of an object is the measured force exerted by that object due to the interaction of that object's intrinsic mass with the local value of gravity; it measures how strongly the force of gravity pulls on that mass. The weight of an object is determined by the operation of weighing that object against another equal force. Given the local value of gravity, an object with a mass of one kilogram will have a weight of 9.8 newton (the SI force or weight obtained by multiplying one kilogram by the acceleration due to gravity of 9.8 m s^{-2}; that is, 9.8 kg m s^{-2} or 9.8 newton) on the surface of the Earth. However, the weight of the same one-kilogram object would be about one-sixth as much on the surface of the Moon, and would be very close to zero when floating freely far out in space away from the influence of the Earth's gravity. The IPK would still have a mass of one kilogram on the Moon, but it would weigh considerably less.

The distinction between mass and weight is unimportant for most practical purposes because the strength of gravity is relatively uniform everywhere on the surface of the Earth. In such a uniform gravitational field, the gravitational force exerted on an object (its weight) is directly proportional to its mass. This means that an object's mass can be measured indirectly by its weight, and so, for everyday purposes, weighing is an acceptable way of measuring mass.

Due to the inevitable accumulation of contaminants on surfaces, the IPK is subject to reversible surface contamination. For this reason, the CIPM declared that, pending further research, the reference mass of the IPK was that immediately after cleaning and washing of the artefact by the specified method.

The old definition of the base unit of mass dates from 1889 and, therefore, predates all modern physics. Thus, we had the extraordinary situation that certain fundamental constants of Nature (the Planck constant, the Newtonian constant of gravitation, the mass of an electron, etc), which are defined to be literally 'universal' were measured in terms of an artefact manufactured in an industrial foundry in the late-19th Century.

We have seen that the move from an artefact-based definition of the metre to a definition based on fundamental physics, the wavelength of visible light has greatly assisted the advance of modern physics; particularly advances made in quantum or atomic physics and nanotechnology. And in May 2019, the kilogram was redefined to the mass corresponding to a particular value of the Planck constant; to be precise $h = 6.626\ 070\ 15 \times 10^{-34}$ J s.

This redefinition of the kilogram will be important for the future development of several fundamental areas of physics; for example, in research concerning the Higgs Boson. This newest member of the menagerie of sub-atomic particles is considered to be the origin of mass; that is, it is the means by which sub-atomic particles couple to the Higgs Field which permeates space to manifest the physical property we term mass. It would have been extraordinary to imagine a determination of the mass of the Higgs Boson in terms of a Victorian object of mass equalling one kilogram. Given the vast difference in masses, such a comparison would like expressing the mass of a small nut in units of the Earth's mass (5.9×10^{24} kg).

The Metric System was first promulgated in April 1795 when the definition of the unit of mass was the mass of one litre of water at a temperature where the density of water is a maximum (about 4 °C). However, the founders of the Metric System knew that this definition was impractical and, therefore, needed to be embodied in a (preferably) small, stable physical object. The necessary artefact was fabricated in 1799 from hot-forged platinum sponge, and is known as the *Kilogram des Archives* (it has been in the French national archives since 1799).

By the time the BIPM was created by the Metre Convention of 1875, there was no longer any thought of basing the definition of the kilogram on the density of water. Rather, the challenge was to replace the *Kilogram des Archives* by another artefact that took full advantage of 19th Century technology, and which would be known as the International Prototype of the Kilogram (IPK).

The IPK was to have a mass that was negligibly different from that of the *Kilogram des Archives*, but would be made of a superior material: an alloy of 90% platinum and 10% iridium by weight. Forty copies of this object were also manufactured; together with the IPK, six copies (or *témoins*) would remain at the BIPM. The majority of the original copies of the IPK became known as national prototypes, and were sent to the various nations who had signed the Metre

Convention of 1875. This system continued to May 2019, except that additional copies of the IPK are manufactured by the BIPM.

Such a copy of the IPK, with a mass of 1 kg ± 1 mg that has been registered with the CIPM was referred to as a 'prototype'. This was the manner in which both the metre and the kilogram were distributed to other nations who wished to adopt the Metric System, when they were both artefact-based units. The metre was redefined in 1982, in terms of the light emitted by a laser, but the kilogram remained an artefact until 2019. An artefact which could only be realized by making a copy of the IPK, comparing the mass of this copy to the IPK and sending that copy to the country wishing to adopt the Metric System together with a certificate of calibration of their copy.

According to the 1886 definition of the kilogram, the mass m_X in kilograms of any object X is given by $m_X = \{m_X/m_K\}$ (kg), where we use the notation of quantity calculus to show the SI unit in square brackets and the number associated with the unit in curly brackets. The comparison of an object with respect to the IPK is carried out by precise weighing, see figure 11.2. Of course, it is not practical to weigh all objects of interest directly against IPK. Instead, an international hierarchy of mass standards was created. From time to time, the national prototypes were brought to the BIPM for comparison with the IPK. It was then the task of each Member State

Figure 11.2. Weighing a 1 kg silicon sphere and a stack of four stainless steel mass standards at the BIPM in exploratory experiments towards the recent redefinition of the kilogram. The measurements are made inside a high-vacuum mass comparator. (This image is reproduced with the permission of the BIPM, which retains full international copyright.)

to disseminate this calibration to those who use secondary mass standards within their borders. Such secondary standards are usually made of non-magnetic stainless steel, with nominal values ranging from below 1 mg to many metric tonnes.

Schematically, this calibration chain can be represented as follows:

$$m_X = \{m_X/m_n\}\{m_n/m_{n-1}\}\ldots\{m_2/m_1\}\{m_1/m_n\}(kg),$$

with every ratio in curly brackets possessing an associated uncertainty. This equation demonstrates how the precision of mass measurements was determined, maintained and promulgated internationally from the IPK; a mass which had, by definition, no uncertainty attached to it. The utility of this expression is founded upon the belief that the platinum–iridium alloy from which the IPK is made is chemically inert and that, therefore, the mass is stable with time. Over the 140 years since the IPK entered into service as the pinnacle of international mass metrology, this initial premise as to the stability of the material, used to fabricate the IPK, was found to be erroneous (see figure 10.2).

11.3 The base unit of electric current is the ampere (A)

A glance at the old definition of the ampere in table 11.1 reveals the impossibility of the experimental realisation of this base unit. Of the seven base units of the old SI, the definition of the ampere, the unit of electricity and the basis of electrical measurements, was by far the most difficult to comprehend. Consider first the experimental geometry described in this definition. It is a perfectly impossible experimental set-up; wires of circular, but negligible, cross-section running to infinity....

What is being considered here is Ampère's Force Law, which says that there exists a force of attraction (or repulsion) between two wires each carrying an electric current (repulsion and attraction depending upon whether or not the currents are flowing in the same direction or in opposite directions, respectively). The physical origin of this force is that the electric current in each wire generates a magnetic field, and the magnetic field generated by the current in one wire exerts a force on the field from the other wire and vice versa. There is thus a force of attraction or repulsion, depending upon the geometry, between the two long wires; as with the polar attraction and repulsion between two permanent magnets (see figure 6.2).

Ampère's Force Law states that the force per unit length, F, between two straight parallel conductors or wires, of constant separation r, is $F = 2k_m (I_1 I_2)/r$; where k_m is the magnetic force constant, and I_1 and I_2 are the currents carried by the two wires. This equation is a good approximation of a solution of Ampère's Force Law when the conductors are of finite length provided that the distance between the wires is small compared to their lengths, but large compared to their diameters; hence the geometry given above in the old definition of the ampere.

The value of k_m depends upon the system of units chosen to define a range of other base units; a force requires the definition of mass, distance and time (see figure 8.2). In the SI, the unit of force is the newton defined as metres multiplied by kilograms per squared second (m kg s^{-2}). Thus, in the old SI the definition of three of the base units of the SI appeared in the definition of the base unit of electric current (see figure 8.2), thereby demonstrating the complex interconnectivity of all the laws of physics. Consequently, a change in the definition of one of those base units had consequences for the definition of other base units. In the SI, $k_m = \mu_0/4\pi$, where μ_0 is the magnetic constant describing how the magnetic field associated with the currents in the two wires propagates through space and was defined in SI units as being equal to $4\pi \times 10^{-7}$ newtons/ampere squared (i.e. $4\pi \times 10^{-7}$ N A^{-2}). Thus, in vacuum the force generated per metre of length between the two parallel conductors carrying current of 1 A and separated by 1 m, is exactly 2×10^{-7} N m^{-1}.

The old, complex definition of the ampere actually served to establish the value of the constant μ_0 as 4×10^7 N A^{-2} exactly. Although the base unit for electricity is the ampere, electrical units were maintained in a practical sense through the volt and the ohm, which were much easier to measure and to realize in a laboratory. Once you have measured the voltage (that is, the electric potential difference, V (volts), given as power divided by electrical current or W/A, which equals m^{2}kg s^{-3}A^{-1}) and the resistance (that is, electrical resistance, R (ohms), given as voltage divided by electrical current or V/A, which equals m^{2}kg s^{-3}A^{-2}) of an electrical circuit, you may use the famous law due to German mathematician and physicist Georg Ohm (1789–1854) to relate V and R to the electric current (I) in the circuit; that is, $V = IR$.

In the early days of electrical metrology, the practical representation of the volt was through batteries, or standard Weston saturated cadmium sulphate electro-chemical cells. The problem was that such Weston cells[1] were temperature-dependent, and they could go flat due to problems with the surfaces of the electrodes, and, consequently, the supposedly constant voltage generated would fall with time.

Since the early-1970s, metrologists have realised the volt by means of the Josephson Effect, a constant and reliable quantum phenomenon. This is a quantum mechanical tunnelling phenomenon discovered by the Welsh physicist Brian Josephson (born 1940) in 1962 when he was a student in Cambridge, and for which Josephson was awarded the Nobel Prize in physics for 1973. A Josephson junction may be formed by sandwiching a thin insulating layer between two superconducting metallic layers. Josephson predicted that if such a junction is driven at frequency f, then its current–voltage (I–V) curve will develop regions of constant voltage at the values $nhf/2e$, where n is an integer and h/e is the ratio of the Planck constant, h to the elementary charge, e. This prediction was verified experimentally, and has become known as the ac Josephson effect. This effect found immediate application in metrology, because it relates the volt to the second through a proportionality involving only fundamental constants. Initially, this led to an improved value of the

---

[1] Named for the English chemist, Edward Weston, the Weston Standard Cell is invariably built in a glass 'H' tube. At the bottom of each 'H', platinum wires are sealed into the glass to make contact with the electrodes of the cell proper. Packed around one wire is liquid mercury, topped with a paste of mercurous sulphate. This forms the positive electrode. Packed around the other wire is a cadmium/mercury amalgam. This forms the negative electrode. The saturated cells would have a layer of cadmium sulphate crystals above the electrodes, and the un-saturated cells would use various packings to retain the electrodes. The top of the cell is then filled with cadmium sulphate solution.

ratio h/e. Today, it is the basis for all primary voltage standards. Josephson's equation for the supercurrent through a superconductive tunnel junction is given by

$$I = I_c \sin\{(4\pi e/h)\int V \ (\mathrm{d}t)\},$$

where I is the junction current, I_c is the critical current, V is the junction voltage. I_c is a function of the junction geometry, the temperature, and any residual magnetic field inside the magnetic shields that are used with voltage standard devices. When a dc voltage is applied across the junction, the current will oscillate at a frequency $f_J = 2 eV \ h^{-1}$; where $2e/h$ is approximately equal to 484 GHz mV^{-1}. The very high frequency and low level of this oscillation make it difficult to observe directly. However, if an alternating current at frequency f is applied to the junction, the junction oscillation at f_J tends to phase lock to the applied frequency, and the average voltage across the junction equals $hf/2e$. This effect is observed as a constant voltage step at $V = hf/2e$ in the voltage–current curve of the junction.

The Josephson effect was initially used to improve the measurement of the constant $2e/h$ based on voltage values derived from the realization of the SI volt, as maintained by Weston chemical cells. The uncertainty of these measurements was limited by the uncertainty of the SI volt realization, and the stability of the Weston cells (see figure 11.3 for details of the evolution of the precision of early voltage standards). The stability of the Josephson volt depends only on the stability of f (which can readily be a part in 10^{12}), and is, at least four orders of magnitude better than the stability of Weston cells. Thus, in the early-1970s, many national standards laboratories adopted a value for the Josephson constant $K_J = 2e/h$ and began using

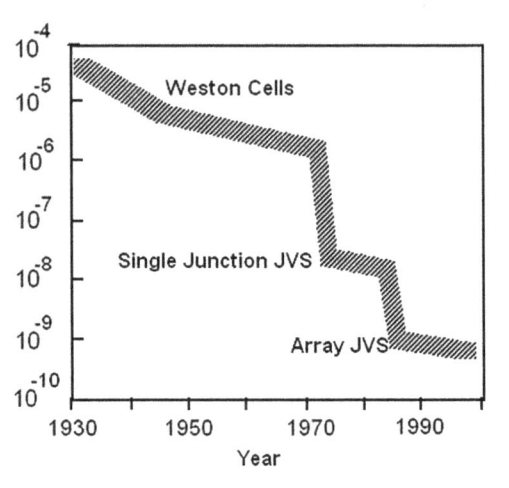

Figure 11.3. The approximate agreement in dc voltage measurements among national metrology laboratories from 1930 to 2000. The vertical axis is the experimental precision of the particular standards. The improvement of the precision, with the introduction of Josephson voltage standards (JVS) is apparent. (Image from https://en.wikipedia.org/wiki/Josephson_voltage_standard#/media/File:Volt_std_agreement_by_year.jpg; it has been obtained by the author from the Wikimedia website where it was made available by EE483597 under a CC BY-SA 3.0 licence. It is included on that basis. It is attributed to EE483597.)

the ac Josephson effect as the practical standard of voltage. Owing, however, to small differences in existing national standards, different values of K_J were adopted; this inconsistency was resolved in 1990 by assigning the constant, $K_{J\text{-}90}$ the value 483 597.9 GHz V^{-1}. What one is observing is a quantum mechanical phenomenon, at low temperatures, between the two superconducting layers. In effect, electrons are tunnelling through the supposedly insulating layer. However, this quantum mechanical tunnelling only occurs at low temperatures and at certain frequencies. The Josephson Effect allows us to relate macroscopic phenomena such as the electrical properties of circuits to fundamental constants of Nature.

The ohm is today realized by another quantum mechanical phenomenon (the quantization of conductivity), the quantum Hall Effect discovered by the German physicist Klaus von Klitzing (born 1943) in 1980, and who was awarded the Nobel Prize for physics in 1985. The quantum Hall effect, or integer quantum Hall effect[2] is a quantum-mechanical version of the Hall effect observed in two-dimensional electron systems subjected to low temperatures and strong magnetic fields, in which the Hall conductance σ undergoes quantum Hall transitions to take on quantized values according to

$$\sigma = I_{\text{channel}} / V_{\text{Hall}} = \nu(e^2/h),$$

where I_{channel} is the channel current, V_{Hall} is the Hall voltage, e is the electronic charge, and h is the Planck constant. The prefactor ν is known as the filling factor, and can take on either integer ($\nu = 1, 2, 3, \ldots$) or fractional ($\nu = \frac{1}{3}, \frac{1}{5}, \ldots$) values. The quantum Hall effect is referred to as the integer or fractional quantum Hall effect depending on whether ν is an integer or fraction, respectively.

The quantization of the Hall conductance permits exceedingly precise measurements. Actual measurements of the Hall conductance have been found to be integer or fractional multiples of e^2/h to, nearly one part in 10^9. It has allowed for the definition of a new practical standard for electrical resistance, based on the resistance quantum given by the von Klitzing constant R_K. The quantum Hall effect also provides an extremely precise independent determination of the fine-structure constant, α, a quantity of central importance in quantum electrodynamics. In a device such as a silicon–metal–oxide–semiconductor field-effect transistor, the Hall voltage V_H for a fixed current I increases in discrete steps as the gate voltage is increased. The Hall resistance, is given by $R_H = V_H/I$ (which is again Ohm's law, but this time related to a quantum mechanical system), and is equal to an integral fraction of the von Klitzing constant, given by $R_K = h/e^2$. In practice, R_K can be measured in terms of a laboratory resistance standard, whose resistance is obtained by comparison with the electrical properties of standard capacitors. In 1990, a fixed conventional value $R_{K\text{-}90} = 25\ 812.807\ \Omega$ was defined for use in resistance calibrations worldwide.

---

[2] The Hall effect is the production of a voltage difference (the Hall voltage) across an electrical conductor, transverse to an electric current in the conductor and to an applied magnetic field perpendicular to the current. It was discovered by American Edwin Hall in 1879.

11.3.1 The Kibble balance

The old definition of the ampere defines a force; a force which arises from the interaction of two electromagnetic fields, but which could be measured. A new device for the determination of the relationship between the mechanical and electromagnetic units is the Kibble balance (originally named in honour of James Watt, but now renamed to acknowledge the ideas of English physicist Bryan Kibble (1938–2016) to use a balancing of forces to make precise measurements). This is a 'balance' which directly compares two forces; that generated by electromagnetic induction, and where a mass is acted upon by the force of gravity.

In its mode of operation, the Kibble balance is a more accurate version of the ampere balance, an early current-measuring instrument in which the force between two current-carrying coils of wire is measured and then used to calculate the magnitude of the current. The Kibble balance operates in the opposite sense; the current in the coils is measured via the Planck constant; thereby measuring mass without recourse to an artefact such as the IPK. Thus, the mass of the object is defined in terms of a current and a voltage—defining an electronic kilogram. The NIST Kibble balance is shown in figure 11.4 (a readable discussion of the NIST

Figure 11.4. The NIST-4 Kibble balance, which began full operation in early-2015, measured the Planck constant to within 13 parts per 10^9 in 2017, which was accurate enough to assist with the 2019 redefinition of the kilogram. The flywheel (balance wheel at the top of the instrument), which is the pivot of the balance has a diameter of 60 cm. The entire device is mounted in a vacuum chamber (image from https://en.wikipedia.org/wiki/Kibble_balance#/media/File:NIST-4_Kibble_balance.jpg; credit: J L Lee/National Institute of Standards and Technology).

device, together with a full list of technical references is to be found at https://en.wikipedia.org/wiki/Kibble_balance) [2].

A conducting wire of length L that carries an electric current I perpendicular to a magnetic field, of strength B experiences a Lorentz force equal to the product of these variables (see table 6.2). In the Kibble balance, the current is varied so that this force counteracts the weight w of a mass m to be measured. This principle is derived from the ampere, or current balance (invented by Lord Kelvin). The force w is given by the mass m multiplied by the local gravitational acceleration g

$$w = mg = BLI$$

The Kibble balance avoids the problems of measuring B and L in a second calibration step. The same wire (in practice, a coil) is moved through the same magnetic field at a known velocity v. By Faraday's law of induction, a potential difference U is generated across the ends of the wire, which equals $B\,L\,v$. Thus

$$U = BLv.$$

The unknown product BL can be eliminated from the equations to give

$$UI = mgv, \text{ and so } m = UI/mv.$$

With U, I, g, and v determined precisely, one obtains an equally precise value for m. Both sides of the equation have the dimensions of power; measured in watts in the SI, hence the original name, watt balance.

The Kibble balance is designed so that the mass under study, and the wire coil are suspended from one arm of a balance, or one side of a fly-wheel—see figure 11.4. The other arm of the balance supports the mass. The system operates by alternating between two modes: 'weighing' and 'moving', see figures 11.5 and 11.6, respectively. The entire mechanical subsystem operates in a vacuum chamber to remove the effects of air buoyancy (see figure 11.4).

While 'weighing' (figure 11.5), the system measures the current, I, and the velocity of the coil, v; the system controls the current in the coil to pull the coil through a magnetic field at a constant velocity, v. Coil position and velocity measurement circuitry are set to an interferometer and a precision clock, to determine the velocity and control the current needed to maintain that velocity. The required current is measured, using an ammeter comprising a Josephson junction voltage standard and an integrating voltmeter. While 'moving' (figure 11.6), the system measures the potential difference, U. The system ceases to provide current to the coil, allowing the counterbalance to pull the coil upward through the magnetic field, which causes a voltage difference across the coil. The velocity circuitry measures the velocity of the coil. This voltage is measured, using the same voltage standard, and integrating voltmeter. A typical Kibble balance measures U, I, and v, but does not measure the local gravitational acceleration, g, because this quantity does not vary rapidly with time. Instead, g is measured in the same location using a precise gravimeter. In addition, the balance depends on a precise frequency reference such as an atomic clock to compute voltage and amperage. The ultimate precision of the mass measurement depends on the Kibble balance, the gravimeter, and the atomic clock.

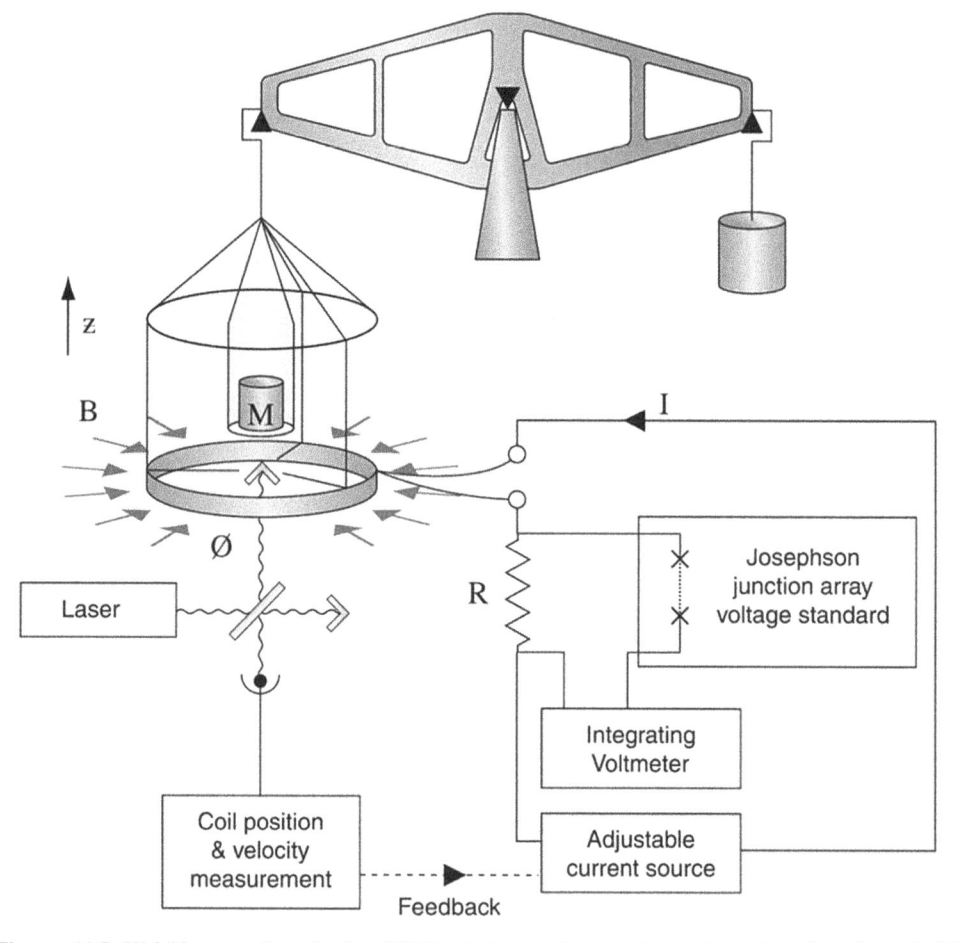

Figure 11.5. Weighing mode of the Kibble balance (image from https://en.wikipedia.org/wiki/Kibble_balance#/media/File:Kibble-balance-in-weighing-mode.png).

Accurate measurements of electric current and potential difference are made in conventional electrical units, which are based on fixed values of the Josephson constant and the von Klitzing constant, $K_{J-90} = 2e/h$ and $R_{K-90} = h/e^2$, respectively. The Kibble balance experiments being undertaken today, are equivalent to measuring the value of the conventional watt in SI units. From the definition of the conventional watt, this is equivalent to measuring the value of the product $K_J{}^2 R_K$ in SI units instead of its fixed value in conventional electrical units:

$$K_J{}^2 R_K = K_{J-90}{}^2 R_{K-90}(mgv/U_{90}I_{90})$$

The importance of such measurements is that they are also a direct measurement of the Planck constant h

$$h = 4/K_J{}^2 R_K.$$

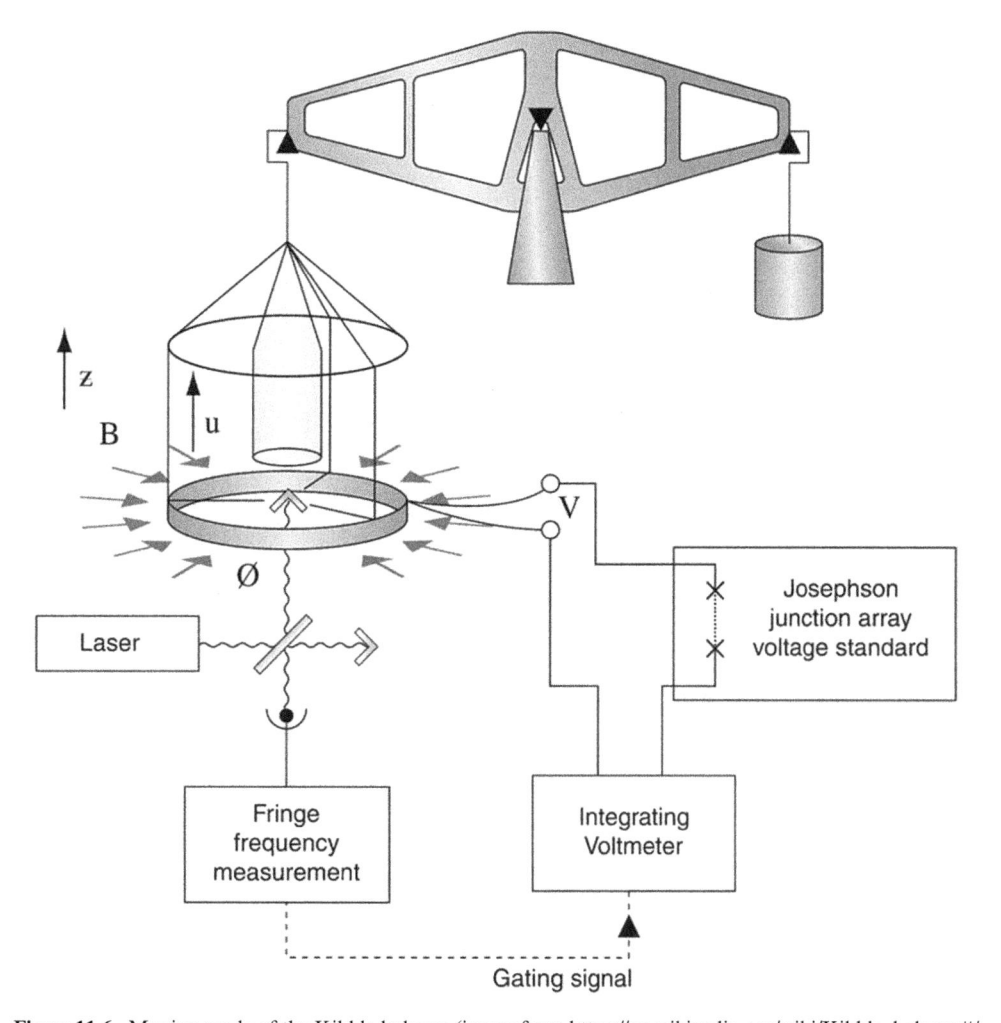

Figure 11.6. Moving mode of the Kibble balance (image from https://en.wikipedia.org/wiki/Kibble_balance#/media/File:The-Kibble-balance-in-moving-mode.png).

The principle of the electronic kilogram relies on the value of the Planck constant, which as of 20 May 2019 has an exact value. This is similar to the metre being defined by the speed of light. With the Planck constant defined exactly, the Kibble balance is not an instrument for measuring the Planck constant, but is instead an instrument to measure mass: $m = UI/gv$.

As can be seen, Kibble balances are large complex devices. Indeed, there is no standard design; each of the 11 balances in use today has unique design features. But they all represent the fulfilment of a longstanding need to redefine the SI unit of mass; to move away from an artefact made in the late-19th Century. Again, the use of quantum mechanical phenomena to define and realise something as large scale or macroscopic as the resistance of a resistor or the current passing through a coil, or a kilogram mass artefact is further evidence that our macroscopic laboratory phenomena are increasing being defined in terms of quantum phenomena.

11.4 The base unit of thermodynamic temperature is the kelvin (K)

The thermodynamic temperature of a material is that material's temperature on the absolute scale of temperature. It is a measure of the fundamental property which underlies temperature; that is, the energy content of the system being investigated. As the temperature of a material falls, the atoms and molecules in the material move more slowly. The zero point, or absolute zero of temperature occurs when the atoms and molecules have no motion, and so cannot reach a lower temperature by being further slowed down. However, this temperature of absolute zero is a temperature which can never be achieved in the laboratory. The old definition of thermodynamic temperature (see table 11.1) meant that the thermodynamic temperature of the triple-point of water was exactly 273.16 K.

This definition of temperature was given by the 10th CGPM of 1954, which selected the triple-point of water as the fundamental fixed point to be used as a reference for all temperature measurements, and assigned to it the temperature 273.16 K. The 13th CGPM (1967/1968) adopted the name kelvin (K), not 'degree Kelvin' or °K, for this unit.

Due to a convention about the manner in which temperature scales have evolved, a thermodynamic temperature, T, is always expressed in terms of its difference from the reference temperature, T_0; for example, $T_0 = 273.15$ K is the ice point or 0 degrees Celsius (where $T_{Celsius} = T_K - 273.15$). The difference between the thermodynamic temperature T and the reference temperature T_0 is termed the Celsius temperature in honour of the Swedish astronomer Anders Celsius (1701–44). Before 1948, the Celsius scale was called the centigrade scale, but this name is no longer used.

The unit of Celsius temperature is the degree Celsius, symbol °C, which is by definition equal in magnitude to the kelvin. So, a difference or interval of temperature may be expressed in kelvin, or in degrees Celsius with the numerical value of the temperature difference being the same. However, the numerical value of a Celsius temperature, expressed in degrees Celsius, is related to the numerical value of the thermodynamic temperature via $T_{Celsius} = T_K - 273.15$.

The unit of temperature in the SI, the kelvin, is named in honour of Lord Kelvin (born William Thomson) and was based on the triple-point of water. This is the temperature at which liquid water, solid ice, and water vapour, the three phases of water, can co-exist in a stable equilibrium; it occurs at 273.16 K (that is, 0.01 °C). At the triple-point, it is possible to change the sample of ice + water + water vapour into ice, into water, or into vapour by making only small changes in pressure and temperature. The fact that water has a triple-point is a manifestation of its unique intermolecular properties when condensed; properties which made life on Earth possible.

In the laboratory, the temperature of water at its triple-point is measured using resistance thermometers; temperature sensors that exploit a well-known and predictable change in electrical resistance of some materials with changing temperature. Resistance thermometers are usually made from platinum, because of the linear relationship between its electrical resistance and temperature, and because of its chemical inertness.

One of the functions of the BIPM is to periodically organize international comparisons of equipment used to realize the base units of the SI. In this way, one can be reassured that, for example, precise measurements of temperature made in one nation may be compared directly with precise measurements of temperature made elsewhere, and that this comparison of measured temperatures can be made to verified levels of precision. Prior to such international comparisons of thermometers, one could not be certain of this comparison of temperature measurements from around the world. Given the importance of having a comparable international system of thermometers monitoring, for example, sea and atmospheric temperature to follow global climate change, such international comparisons are essential for obtaining accurate data for the climate scientists to use in their models and predictions.

An international comparison of the triple-point of water cells undertaken at the BIPM was a measure of the world's ability to precisely measure the same subtle changes of temperature. Such an international comparison of triple point cells from 12 nations, carried out by the BIPM from 1994 to 1996 demonstrated that national capabilities in temperature measurement agreed to within ±0.1 mK (0.0001 kelvin or 0.0001 °C). In a more recent comparison, the final level of precision had increased, with a measured uncertainty of ±13 µK. This impressive improvement demonstrated that the technology for measuring temperatures had evolved significantly between the two international comparisons.

What this increased level of precision also demonstrated was that it was now possible to identify the isotopic composition of the various water samples sent to the BIPM with the triple-point cells. That is, the measurements made at the BIPM were sufficiently precise to be able to directly compare the properties of the different water samples. Properties of the sample that depend on the number of molecules in a given volume of solvent and not on the properties, size or mass of a sample composed of a single molecule.

The Law of Colligative Properties was established by the French chemist François-Marie Raoult and tells us how the freezing point and the boiling point of a liquid are functions of the composition of that liquid. For example, pure water will freeze at 0 °C and boil at 100 °C, while sea water will freeze at a temperature lower than 0 °C and boil at a temperature higher that 100 °C. But even if one repeatedly distilled a sample of water, we cannot make it isotopically pure. That is, in a sample of distilled water, all of the molecules will be H_2O, but there will be some heavy water or water with hydrogen replaced by its heavier isotope, deuterium (H-2), which has a natural abundance of about 0.0026%. In addition, the most abundant isotope of oxygen is O with mass 16 (O-16), but other isotopes of oxygen are found naturally. Thus, supposedly pure water, will not be isotopically pure, and this mixture of isotopes can, and does, affect the measurements of the resistance of water triple point cells and hence the realization of the kelvin.

The isotopic composition of oxygen atoms in the Earth's atmosphere is 99.759% O-16, 0.037% O-17 and 0.204% O-18. Because water molecules containing the lighter isotope are slightly more likely to evaporate and fall as precipitation, fresh water and polar ice on Earth contains slightly less (0.1981%) of the heavy isotope O-18 than air (0.204%) or seawater containing (0.1995%). This disparity permits a study of the evolution of global temperature patterns via analysis of ice cores.

In the most recent BIPM comparison of water triple-point cells, the cells of those laboratories which used water of the isotopic composition of ocean water were clearly identifiable from the results from cells containing isotopically pure water. Consequently, the two sets of cells realized slightly different temperatures. As a result of international discussions, it was necessary to change the definition of the kelvin to better define the isotopic composition of the samples under investigation. This makes the realization more difficult, but the change was necessitated because of the advances made in the technology of making precision measurements of resistance.

In science it is often the case that the more detailed your investigation of even a well-known phenomenon become, the greater the derived insight; and often new phenomena are discovered. Consequently, at its 2005 meeting the CIPM amended the definition of the kelvin by affirming that the well-known definition refers to water having the isotopic composition defined exactly by the following isotope compositions: 0.000 155 76 *mole of H-2 per mole of H−1*, 0.000 379 9 *mole of O-17 per mole of O-16, and* 0.002 005 2 *mole of O-18 per mole of O-16.*

The old definition of thermodynamic temperature provided a solid basis for temperature measurements over certain ranges, but it left temperature and a variety of other thermal quantities slightly separated from the rest of the SI (see figure 8.2). But the temperature of a system is not a fundamental physical property of a system. The temperature of a sample of water, for example, is related to the energy content of that sample, which is the more fundamental physical property through the Boltzmann constant (k_B), which is the constant of Nature relating the energy content of a bulk system with the energy content of the atoms and molecules which compose the bulk. Since May 2019, the definition of thermodynamic temperature has involved a fixed value of the Boltzmann constant (see table 11.1) [3].

11.4.1 The Boltzmann constant

The Boltzmann constant, k_B, is the conversion factor between mechanical energy and temperature, it tells us how many joules of energy we must give to a molecule in order to increase its temperature by one kelvin (or one degree Celsius).

Ludwig Boltzmann's grave in Vienna is marked by a monument upon which the following formula is carved (often referred to as Boltzmann's equation)

$$S = k \ln W. \tag{11.1}$$

Here S is entropy, W represent the nature of the state being investigated, and the constant k (today written as k_B) is Boltzmann's constant (the best value available today is $k_B = 1.380\ 649 \times 10^{-23}$ J K^{-1}). This constant of Nature relates the energy content of a single molecules to the energy content (temperature) of a bulk sample of that molecule via Avogadro's number; that is, $N_A k_B = R$, the gas constant. Interestingly, equation (11.1) was never written down by Boltzmann (1844–1906), but was derived by Max Planck (1858–1947), who also first defined the Boltzmann constant. Planck came to the Boltzmann constant through his study of the nature of the radiation emitted by a hot black-body; a study which also allowed him to lay the foundations of quantum mechanics.

Max Planck was interested in the energy radiated by a hot black-body per unit volume, in a frequency range ν to $\nu + d\nu$ at a temperature T. In 1900, he found that the radiative energy, $\rho(\nu, T)$ is given by

$$\rho(\nu, T) = (8\pi h\nu^3/c^3)\{1/\exp h\nu/k_B T - 1\}, \tag{11.2}$$

where h is the Planck constant and k_B is the Boltzmann constant. Planck, not unnaturally wished to explore the consequences of this equation, and it is this examination which made him the discoverer of quantum mechanics [4].

A modification of equation (11.1) is found for the first time in a paper by Planck of 1900 [5]

$$S = k \ln W + \text{constant} \tag{11.3}$$

The essence of equation (11.3) [and equation (11.1)], the discovery that the Second Law of Thermodynamics can be understood only in terms of the connection between entropy and probability is one of the cornerstones of modern physics and engineering. The First Law of Thermodynamics, in the form of energy conservation is probably of great antiquity, but in its more modern form (the impossibility of achieving perpetual motion of the first kind) dates from the period 1830 to 1850. The Second Law of Thermodynamics was discovered by Rudolf Julius Clausius while he was investigating the work on heat engines by Nicolas Léonard Sadi Carnot (1796–1832). In its original form, the second law said that heat cannot go from a colder to a warmer body without some other accompanying change. The term entropy was first coined by Clausius in 1865, when he stated the two laws of thermodynamics as, *The energy of the world is constant, its entropy strives to a maximum*, and commented that 'the second law is much harder to grasp than the first law.' It was, however, Maxwell who first stated that the second law is statistical in nature. In 1872, while responding to a question about the nature of Maxwell's demons, he said, 'The problems of the mechanical theory of heat are…problems in the theory of probability.'

The Boltzmann constant is one of the constants of Nature, whose values have been fixed by the new definitions of the SI (see table 11.1). Below we outline some of the steps in this 'fixing' of the value of something that has always previously been a measurable quantity. Before May, 2019 we measured temperatures in terms of other temperatures. Indeed, we measured how much hotter or colder something was relative to a particular target temperature; the triple-point of water. Since May 2019, the best estimate of the Boltzmann constant, based on numerous measurements from around the world has been established and fixed to more closely define temperature in terms of the internal energy of molecules. This represents a fundamental change in our conception of the unit of temperature and of what we mean by 'one degree'.

For the last decade or so, international research teams have been making measurements of the Boltzmann constant, using a variety of techniques [3]. The aim of this endeavour has been to make measurements with as low a measurement uncertainty as possible. Every four years, the organisation CODATA[3] critically reviews all the published estimates of fundamental constants made in the previous four years, and generates a set of recommended values. These CODATA recommendations are a 'weighted' average of the published data giving more weight to estimates which have a low measurement uncertainty; this weighting is done via a least-squares fitting procedure[3].

Today, a new measure of the link between temperature and molecular energy has been established, which will be reflected as a change in our temperature scale, not a change in the Boltzmann constant, which is now fixed. Fixed until such time as there are significant paradigm changes in how we regard the constants of Nature and the partitioning of available thermal energy in the many degrees of freedom that exist in all molecules, and in condensed phases make of those molecules.

11.5 The base unit of light intensity is the candela (cd)

Prior to 1948, various standards for luminous intensity were in use in a number of countries. These were typically based on the brightness of the flame from a 'standard candle' of defined composition, or the brightness of an incandescent filament of specific design. One of the best-known of these was the English standard of candlepower; one candlepower was the light produced by a pure spermaceti candle weighing one sixth of a pound and burning at a rate of 120 grains per hour. This 'standard candle' was made from the waxy substance found in the head cavities of the sperm whale. Spermaceti is created in the spermaceti organ inside the whale's head. This organ may contain as much as 1900 litres (500 US gal) of spermaceti.

The old definition of the candela (see table 11.1) is a statement about the amount of light, of a particular frequency measured in a certain solid angle. The frequency corresponds to light with a wavelength of 555 nm, in the green area of the spectrum where the human eye has evolved to be most sensitive. This definition of the candela indicates its age; it is referring to an epoch when measurements of light intensity were made with the human eye, which is actually a sensitive measuring device having evolved to detect as few as two green photons. The green light mentioned in the

---

[3] The Committee on Data for Science and Technology (CODATA) was established in 1966 as an interdisciplinary committee of the International Council for Science. CODATA is best known for its Task Group on Fundamental Constants. Established in 1969, its purpose is to periodically provide the international scientific and technological communities with an internationally accepted set of values of the fundamental physical constants and closely related conversion factors for use worldwide. The first such CODATA set was published in 1973. The latest version is Version 7.0 called '2014 CODATA' published on 25 June 2015. The CODATA recommended values of fundamental physical constants are published at the NIST Reference on Constants, Units, and Uncertainty. Their website is: http://www.codata.org/, and is the primary source for the most precise and up-to-date values of physical constants.

definition is assumed to be emitted, by an appropriate source in every direction; that is, on the surface of a sphere, and the definition refers to an amount (the watt is the radiant energy per unit time) emitted into a particular small angle (a steradian is defined as the solid angle subtended at the centre of a sphere of radius *r* by a portion of the surface of the sphere), which is seen by the detector.

Of particular importance to us, is the unit of light intensity based on measurements made of the light emitted by flame and incandescent filament sources. With the adoption of the unit of the 'new candle', promulgated by the International Committee for Weights and Measures (CIPM) in 1946, the definition of the candela evolved rapidly. In early-1948, the definition of the new candle was: 'The value of the new candle is such that the brightness of the full radiator at the temperature of solidification of platinum is 60 new candles per square centimetre.' The name, candela was then ratified in 1948 by the 9th CGPM. In 1967, the 13th CGPM removed the term 'new candle' and gave an amended version of the candela definition: 'The candela is the luminous intensity, in the perpendicular direction, of a surface of 1/600 000 square metre of a black-body at the temperature of freezing platinum under a pressure of 101 325 newtons per square metre.'

A Planck radiator or black-body is an idealized solid that absorbs all radiation incident upon it (hence its name). Because of this perfect absorptivity at all wavelengths, a black-body is also the best possible emitter of thermal radiation, which it radiates in a characteristic, continuous spectrum that depends only on the temperature of the black-body. Something of the physics of black-bodies is given above in the discussion of the Boltzmann constant; they have a hallowed place in the history of physics, as it was by the study of the radiation emitted by black-bodies that Max Planck invented quantum mechanics[4]. The thermal radiation emitted by a black-body arises from the mechanical energy of the vibrations of the constituent atoms. The faster the atoms vibrate, the hotter is the material and the greater the amount of radiation emitted, and the wider the range of frequencies emitted by the Planck radiator. This emitted radiation is called black-body radiation and has a distribution of frequencies with a characteristic (Maxwell-Boltzmann) form. As the temperature increases past a few hundred degrees Celsius, black bodies start to emit light at visible wavelengths, appearing red, orange, yellow, white, and blue with increasing temperature. When an object, which is being heated is visually white (that is, 'white hot'), it is actually emitting a substantial amount of ultraviolet and infrared radiation.

It was through his study of the radiation emitted by black-bodies that Planck first conceived of the quantum theory. Planck's work on this radiation was based on the research of the German physicist Gustav Robert Kirchhoff (1824–87), who first described the mathematical properties of the ideal black-body, a term which he invented. Kirchhoff could therefore be said to be the grandfather of the quantum theory.

---

[4] Does one invent or discover something as fundamental as quantum mechanics—a fascinating question, but not germane to our present discussion.

Unfortunately, given the mathematical form of the law which governs the intensity of radiation, at a particular wavelength emitted by a black-body at a particular temperature, it is not at all straightforward to make measurements of the light intensity emitted into a solid angle. Indeed, it is always a range of wavelengths and not a single monochromatic wavelength which has to be treated.

Given that detailed modelling and measurement of the radiation emitted by black-bodies is complex, and often leads to results which are not very precise, there is more uncertainty associated with the realization of the base unit of light intensity than with the other base units of the SI[5]. Interestingly, the procedure of determining light intensity is based on the light emitted by a hot object, and there are many scientists today who question the utility of the concept of light intensity as one of the base units of the SI. After all, what is really at issue is the temperature of the black-body, and temperature is already one of the base units of the SI. Consequently, there are some scientists who say that there need only be six base units in the SI. Light intensity was something of a distraction during earlier stages of the development of the scientific aspects of the Metric System (that is, before the advent of quantum mechanics which allows one to consider the energy content of electromagnetic waves). The nature of the light emitted by a black-body depends upon its temperature, which is already defined in the SI, and the radiant flux or amount of light emitted by the black-body can also be called power ($M L^2 T^{-3}$), which is already defined in the SI as the rate at which work ($M L^2 T^{-2}$) is performed. The dimension of power is energy divided by time. The unit of power is the watt (W), which is equal to one joule per second. Hence, the day may not be too far away when there is a move to reform the SI by going from seven base units to six base units, by the suppression of the candela, which came into being before the nature of light, energy and the nature of thermal radiation in solids, and how they are interrelated had been fully comprehended.

11.6 The base unit of amount of substance is the mole (mol)

The old and the new definition of the unit of quantity of something are given in table 11.1. When the mole is being discussed, the elementary entities must be specified; they may be atoms, molecules, ions, electrons, or other particles.

The old definition was adopted by the 14th CGPM of 1971, after consultation with the international community of chemists. Thus, chemistry came late into the SI, and one naturally asks the question, why was this the case? Well, the truth is that up until the beginning of the 20th Century, physicists did not believe in the reality of molecules [4].

Indeed, it was not until after the physics community had accepted Rutherford's famous atomic model of 1913 where the atom is a dense, compact nucleus with orbiting electrons that they started to take seriously the necessity of explaining the chemical changes that chemists had been observing, investigating and recording

---

[5] This can be seen in table 11.1, where the luminous efficacy of radiation, at a frequency of 540×10^{12} Hz (550 nm) is only moderately precise at 683 lm W^{-1}—a part in a thousand.

since the days of the Renaissance alchemists [molecular meme]. Chemists may have been convinced of the existence of atoms, but physicists needed more convincing; and it was in 1925 with the arrival of Eugene Wigner's spherical model of the atom and of Wolfgang Pauli's model of how electrons are coupled in atoms (his exclusion principle) that physicists finally accepted the existence of individual atoms, with internal structure [4].

Although the concept of the atomic structure of matter was widely accepted by the late-18th Century, the two millennia which separated the atomists (Democritus and Epicurus) and the late-18th Century *savants* who had an interest in chemical transformation (for example, Antoine-Laurent de Lavoisier) had changed very little about our understanding of the structure of everyday matter, and what was considered to be the primal matter from which the Universe was believed to be composed.

For the gifted chemist, the ability to convert molecule A into molecule B is as much an art as it is a science. Even though the experimental chemist may not know any electromagnetism, he is able to manipulate matter to effect chemical transformations. By the time that physicists started to see beyond their continuum models for the nature of matter, and accept what chemists had been saying for generations, the other base units of the SI had been established. But it still took until 1971 for the mole to be fully integrated into the complex network of coupled quantities that is the SI; see figure 8.1.

It was the English chemist, John Dalton (1766–1844), who started examining the composition of the things he found around him, and began to identify a finite number of atomic species, or chemical elements. During Dalton's life, 18 chemical elements had been identified, and it is certain that Lavoisier would have made significant contributions to chemistry had he not lost his head in 1794. lavoisier's genius was expressed by his colleague from the Commission of the *Académie des sciences* which invented the Metric System, J-F Lagrange, who lamented his friend by saying: '*It took them only an instant to cut off his head, but France may not produce another such head in a century.*'

The first chemical analyses gave the ratios of the elements that compose molecules, and early-chemists invented the stoichiometric chemical symbols that we still use today to write chemical reactions in shorthand, and to assist in investigating chemical change. By investigating the quantities and proportions of, for example, oxygen and carbon that came together to form carbon dioxide, these early chemists determined how many atoms of X would react with how many atoms of Y. However, even in 1860, chemists could still talk of the difference between physical and chemical molecules, where a chemical molecule was the smallest entity capable of participating in a chemical reaction, and a physical molecule was the smallest entity that composed a gas, a liquid or a solid. There were heated international debates about the differences between physical and chemical molecules. All of this was only resolved with the advent of quantum mechanics in the early-20th Century [4 (chapter 8)].

The principle question was whether atoms were real objects or only mnemonic devices for coding chemical structures and stoichiometric ratios. For the physicist, the structure of the smallest entity in the sample of a gas, a liquid or a solid was unimportant, what mattered was how it transmitted energy (the continuum model). For the chemist, it was the ratio of different atoms in the reacting species which reacted to generate new compounds, that was important. But how do you get to the essence of chemical change? The problem was resolved by an Italian chemist Lorenzo Romano Amedeo Carlo Bernadette Avogadro di Quaregna e Cerreto, Count of Quaregna and Cerreto (1776–1856), who formulated what we today call Avogadro's law, which allows chemists to quantitatively determine how much carbon, hydrogen and oxygen react to produce, for example, glucose. The principle is based upon the atomic weight of each element; the atomic weight of carbon is 12, the atomic weight of oxygen is 16, and for hydrogen it is 1.

Avogadro's ideas allowed chemists to look at the relative amounts of reacting elements by establishing a quantitative basis for the amounts of X and Y which react to give Z. Avogadro's law says that there is an amount or quantity of each atom or molecule, called the Avogadro number, N_A, which is another fundamental constant of nature and is, in fact, a very large number (about 6×10^{23}) which represents one mole of that atom or molecule. Avogadro explained the numbers in chemical formulae and moved chemistry from a qualitative basis to a quantitative basis; chemistry was no longer a 'black art' ruled by hit-or-miss laboratory practice.

Physicists and chemists agreed to assign the value 12 (exactly) to the atomic weight of the isotope of carbon with mass number 12 (carbon-12), and the atomic weight of other atoms are determined relative to this fundamental atomic weight. This unified scale gives the relative atomic and molecular masses, also known as the atomic and molecular weights, respectively.

It follows from the old definition of the mole (see table 11.1) that the molar mass of carbon-12 is exactly 12 grams per mole (12 g mol^{-1}). In 1980 the CIPM approved the unrealizable specification that: 'In this definition, one is referring to the unphysical and unrealizable situation of unbound or isolated atoms of carbon-12, at rest and in their ground state', at that time there was no way to realize the mole as it would have required the measurement of precisely Avogadro's number of carbon atoms, which although only 12 grams would still have required an unavailable level of precision. We will see in chapter 13 how values of Avogadro's number (N_A) are measured. A history of the measurement of N_A is to be found in [4 and 5].

The new definition of the mole determines the value of the constant of Nature that relates the number of entities to amount of substance for any material. This is Avogadro's constant, N_A. If $N(X)$ denotes the number of entities X in a specified sample, and if $n(X)$ denotes the amount of substance of entities X in the same sample, we may write $n(X) = N(X)/N_A$. Note that since $N(X)$ is dimensionless, and n (X) has the SI unit mole, the Avogadro constant has the coherent SI unit reciprocal mole. This relationship is the basis of all chemistry.

Further reading

[1] Concerning the evolution of the SI since the late-19th Century, the best source is the bilingual SI Brochure published by the Bureau international des poids et mesures (BIPM). This is a non-technical document intended for those already familiar with the science relating to the origin of the SI; and is essentially a list of rules concerning the use of SI units. The brochure is written by the Consultative Committee for Units of the BIPM; the most recent edition, the ninth was published in 2019. This substantial booklet is only available through the BIPM, but the text is freely available on the BIPM's website (https://bipm.org/en/publications/si-brochure/). Also included are extensive lists of references as to when certain words or quantities were adopted for use with the SI. The BIPM website contains many useful documents detailing the history and characteristics of the base units of the SI.

[2] Robinson I A and Schlamminger S 2016 The watt or Kibble balance: A technique for implementing the new SI definition of the unit of mass *Metrologia* **53** A46–74

[3] Fischer J *et al* 2018 The Boltzmann project *Metrologia* **55** R1–20

[4] Williams J H 2018 *The Molecule as Meme* (San Rafael, CA: Morgan & Claypool Publishers)

[5] Pais A 1982 The light quantum *Subtle is the Lord: The science and the life of Albert Einstein* (Oxford: Oxford University Press) chapter 19

Chapter 12

The base units of the Système International des Unites (II)

In matters related to the SI, there is a traditional order in which the seven base units are discussed. I have deliberately separated the discussion of the second from the other base units, as the second is likely to be redefined within the near future. Consequently, this chapter is a discussion of modern trends in time metrology.

12.1 The base unit of time is the second (s)

The caesium atoms, mentioned in the definition of the second in table 11.1, are contained in what is termed an atomic clock. It follows from this definition that the frequency of the hyperfine splitting in the ground state of the nucleus of an atom of a particular isotope of caesium (caesium-133) is exactly 9 192 631 770 Hz (s^{-1}). This definition was further specified in 1997, when the CIPM added the unphysical and unachievable *caveat* that: *This definition refers to a caesium atom at rest at a temperature of 0 K*; that is, at the absolute zero of temperature.

Prior to quantum mechanics, the fundamental unit of time, the second, was simply considered to be the fraction 1/86 400 of the mean solar day. The exact definition of the mean solar day was left to astronomers. The first astronomers to investigate the nature of the second lived in Ancient Sumeria over five millennia ago, and they were considerable astronomers. The history of astronomy begins with the ancient Sumerians, who developed the earliest writing system about 3500–3200 BCE.

The earliest Babylonian star catalogues date from about 1200 BCE. The fact that many of the stars named in these Babylonian catalogues appear in Sumerian rather than the language of Ancient Babylon suggests a continuous study of the night sky reaching back into the early-Bronze Age. Among other things, these ancient astronomers used a number system to the base 60 (sexagesimal), which simplified the task of recording very large and very small numbers, and they invented the angular view of the sky, which we still use today; the sky as a flat circular disc of 360° divided up into minutes and seconds of arc, with 60 s in a minute and 60 min in a degree. The ancient Sumerians were the first to mix mythology and astronomy and divided the 360° of the night sky into the Houses of the Zodiac and gave us the familiar pictures of the night sky which are still part of our culture.

The earliest definition of the second dates from the first Sumerian cities, well before 3000 BCE, making the second the oldest unit of measurement still in use. But why did these Bronze-Age people consider the second to be the most fundamental unit of time, when they clearly saw that the Earth rotated around the Sun in more than 365 of the units which defined the rotation of the Earth about its axis (the day)? And that the Lunar cycle was about 29 of these units of the rotation of the Earth about its axis? They divided the day into shorter units, which today we call hours, the hours into minutes and the minutes into seconds; but why did this division of time stop here? Perhaps the reason is that those who were interested in this subject: the magicians, doctors and priests looked at the human body and realized that the rate at which our dependable internal clocks beat is about 60–70 beats to the minute, that is, about 1 s. And these ancient people realized that man did not need a shorter time standard, as his own heartbeat to the fundamental frequency of life, which they were then able to relate to the movement of the stars and planets.

This holistic picture of man's place in the Cosmos has now been lost, but it remained central to a view of Nature up until the Enlightenment. But time is a strange quantity. We feel it pass. We age, we grow old and so we know time exists. But unlike a fundamental unit of distance, or mass, or electricity, or temperature, we have no concept of an absolute, ideal time. All our clocks measure the same thing, but is it time or has time been defined to be what our clocks measure?

Over the last few centuries, increasingly precise astronomical measurements have demonstrated that irregularities in the rotation of the Earth made Babylonian astronomy an unsatisfactory definition for the second. Due to tidal forces on the surface of the Earth produced by the Moon acting via gravity on our seas and oceans, the length of the day is increasing by about 1.4 ms per century (which mounts up after many millennia); the Earth is slowing down.

In order to define the base unit of time more precisely, the 11th CGPM of 1960 adopted a definition of the second provided by the International Astronomical Union, where the second was standardized to the duration of one particular year, measured on one part of the globe; to be precise, the tropical year 1900. However, even then scientists (spectroscopists) were demonstrating that an atomic standard of time, based on a transition between two energy levels of an atom, or a molecule could be realized and reproduced much more accurately and, more importantly anywhere on Earth. And it was for this reason that the 13th CGPM of 1967 and 1968 adopted the definition of the second given in table 11.1.

12.1.1 Atomic time

Traditional time-keeping devices are mechanical, usually circular in appearance, employing cogs, wheels and pendulums as the means of beating seconds, and then using 'hands' or needles moving over a circular, inscribed 'clock face' to record the accumulated passage of those seconds (see figure 4.1). An atomic clock, however, is a clock that employs the measurement of the frequency of microwave radiation used to excite a transition in the nucleus of a particular atom for its timekeeping element. In an atomic clock, time is defined as the difference between two energy levels of an atomic nucleus; this approach to time keeping is only possible through the application of quantum mechanics.

Under ambient conditions, the atom of interest is in the lower of two energy states, E_1 and E_2. From this state, it may be excited (see figure 12.1) to the upper state by absorption of a quantum of energy (a photon) of the appropriate energy (or via Plank's equation, $E = h\nu$, frequency). The electron in E_1 absorbs a photon of light and is excited to the higher energy state, E_2. Depending upon the lifetime of this excited state, the atom will then emit a photon (fluorescence) to return to the ground state, E_1. As can be seen, the development of quantum mechanics is intimately linked with the development of atomic (and molecular) spectroscopy. Indeed, spectroscopy could not have been established via classical mechanics.

Atomic clocks are among the most accurate frequency standards known and are used as the primary standards for the international standardization of time in, for example, controlling the frequency of television and radio broadcasts, global navigation satellite systems such as GPS, and the interfacing and networking of computer systems worldwide.

An atomic clock contains either hydrogen gas or the vapour of a metal such as caesium. These two chemical elements are particular in that the electron (in the case of hydrogen atom) and the electrons (in the case of caesium atom) interact with the nucleus of the atom to create two possible orientations of the nucleus in the atom. These two nuclear states co-exist in equal concentrations in the vapour of the element at room temperature, but when the temperature of the vapour is cooled to close to absolute zero, one (that of lowest energy) of the particular nuclear states dominates. This dominant low-temperature nuclear orientation may then be excited by microwave radiation into the other possible nuclear orientation. Thus, at low temperatures these vapours are seen to absorb microwave radiation at a very precise frequency.

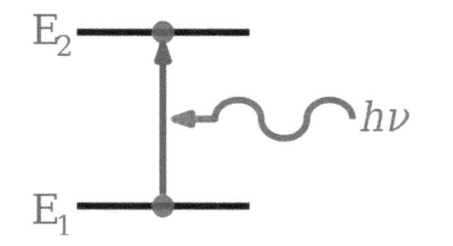

Figure 12.1. The schematic of an atomic absorbtion.

The first true atomic clock was based on caesium and was built by Louis Essen (1908–1997) in 1955 at the National Physical Laboratory, UK[1]. Calibration of the caesium standard atomic clock was carried out by the use of the astronomical time scale ephemeris time. This definition of the second as a combination of atomic physics and astronomical measurements was based on observations of the Moon, whose ephemeris is tied indirectly to the apparent motion of the Sun, and was equivalent to the previous definition within the limits of experimental uncertainty. What we have here, is another use of a quantum phenomenon, the absorption of microwave radiation by the atomic nucleus of a caesium atom, to measure and characterize an astronomical phenomenon. The ability to combine measurements and observations of the very small with measurements and observations of the very large is unique in physics, where it has still not been possible to formally unify atomic-based quantum mechanics and astronomical celestial mechanics.

National metrology laboratories routinely maintain, in their atomic clocks an accuracy of 10^{-9} s per day, or about a part in 10^{14}; a precision of one second in 3 million years; such high levels of precision are unknown in other areas of science (see table 12.1 for details). These clocks collectively define a continuous and stable time scale, International Atomic Time; abbreviated as TAI from the official French name, *Temps atomique international*. For civil time, another time scale is calculated and disseminated, Coordinated Universal Time; UTC from *Temps universel coordonné*. UTC is derived from TAI by the *Bureau international des poids et mesures* (BIPM).

In 1884, the International Meridian Conference, Washington DC, USA, chose local time at the Royal Observatory, Greenwich, UK, as the world's standard of time leading to the widespread use of Greenwich Mean Time (GMT). In 1929,

Table 12.1. Metrological properties of various types of frequency standards (clocks based on atomic spectroscopy).

Type of clock (carrier of the spectrum)	Absorption frequency in Hz	Relative Allan deviation[a]
^{133}Cs	9 192 631 770	10^{-13}
^{87}Rb	6 834 682 610.904 324	10^{-12}
^{1}H	1 420 405 751.7667	10^{-15}
Optical clock (^{87}Sr)	429 228 004 229 873.4	10^{-17}

[a] Allan deviation is a statistic used to estimate stability; it is the Allan variance, but since it is the square root of the variance, its proper name is the Allan deviation. An Allan deviation graph typically shows the stability of the device improving as the averaging period of its output gets longer—since some types of noise can be removed by averaging. At some point, however, more averaging no longer improves the result. This point is called the noise floor; the point where the remaining noise consists of nonstationary processes such as aging, or random walk.

---

[1] Louis Essen was often called the Time Lord. Well, his work on time metrology came to fruition during the 1960s in the NPL, Teddington, when the best-known children's science fiction series on TV was Dr Who. However, Essen's research dates back to the Second World War.

Universal Time (UT) was introduced to refer GMT to the day starting at midnight. UT is a time-scale based on the rotation of the Earth, and provides the mean solar time on the meridian at Greenwich, UK, which is the zero meridian of geographic longitude.

The advent of the atomic clock in 1955 provided a form of time-keeping that was both stabler and more convenient than astronomical observations; and TAI began to be widely used in 1958. While it was clear that basing measurements of time on unchanging atomic properties would produce advances in time-keeping, it was accepted that civil-time should be kept synchronized with the culturally familiar rotation of the Earth; for example, navigation relies on UT. Humanity has an experience of the passage of time in terms of the rotation of our planet about its axis, and the rotation of our planet about the Sun; the scientific community might wish to adopt atomic time, but humanity needs to keep its culturally familiar units of time. There was thus a need to modify the atomic time scale and relate it to UT.

Coordinated Universal Time (UTC) is a time scale derived from TAI; both scales have the same unit (the atomic, SI second). To allow UTC to follow as closely as possible Universal Time, leap seconds are added to UTC at an irregular interval. These leap seconds are added to compensate for the Earth's slowing rotation. The leap seconds allow UTC to closely track UT, which is a time scale based on the Earth's rotation, rather than a uniform passage of seconds as derived from atomic clocks. The International Earth Rotation and Reference Systems Service monitors the irregularities of the Earth's rotation and is responsible for taking decisions about the need for applying leap seconds to UTC. To date, UTC and TAI differ by about 33 s, as a result of the addition of leap seconds.

Even though atomic clocks are remarkable technological devices, which have been responsible for facilitating modern communications, permitting global capitalism (the global village has a single atomic time-piece), the Internet and space travel, there remains an enormous divide when it comes to the comprehension and understanding of atomic time. There is after all, no cultural link between the oscillations of atomic nuclei at low temperatures, and the all-present sensation of the rotation of the Earth about its axis, which defines our days and nights; and the rotation of the Earth about the Sun, which defines our seasons and years. The world and capitalism may run according to UTC, and you can see the numbers corresponding to UTC displayed when you make an electronic bank transfer, but as it has no cultural point of reference compared with the millennia of familiarity of Babylonian time, it may never be fully comprehended, or accepted by everyone.

12.1.2 High-resolution atomic spectroscopy and time metrology

At the beginning of the last century, physicists did not believe in the existence of the molecules that chemists had been manipulating for centuries. It was the advent of modern quantum mechanics in the late-1920s that showed physicists how electrons were coupled together in the supposed indivisible atom, and which finally allowed the molecule meme to make its inexorable progress in converting physicists to the reality of molecules.

The first attempts at seeking to understand the internal workings of the atom focused on atomic line spectra. Such spectra are observed in the light emitted by atomic discharge lamps, where a current is passed through a low-pressure gas, and the atoms are excited or even ionized and relax to the ground state by emitting light. This emitted light is observed as a series of sharp lines corresponding to electric-dipole allowed transitions between the electronic states of the atom. The Danish physicist, Niels Bohr (1885–1962), famously unravelled the main features of the atomic line spectrum of the hydrogen atom using the original quantum theory of Planck. Indeed, this was the first use of Planck's quantum theory outside of thermodynamics. And the developing science of spectroscopy went hand in hand with the development of quantum mechanics. So, it is perhaps fitting that the modern basis of time metrology, should be based on the quantum mechanical model of multi-electron atoms.

However, Bohr's model of the hydrogen atom could not be made to fit the next most complex atom (helium with two electrons), or indeed, explain the high-resolution features of the hydrogen spectrum. In the case of helium, the problem was to identify where the two electrons were in the atom, and how were they coupled? The Dutch physicists George Uhlenbeck (1900–1988) and Samuel Goudsmit (1902–1978) realized in 1925 that the description of atomic spectra could be simplified if it was assumed that an electron possessed an intrinsic angular momentum with quantum number $s = \frac{1}{2}$, and which could exist in two states; with $m_s = +\frac{1}{2}$, denoted α or $\uparrow$, and $m_s = -\frac{1}{2}$, denoted β or $\downarrow$. This intrinsic angular momentum is called the spin of the electron. The spin of sub-atomic particles is a purely quantum mechanical phenomenon in the sense that when h tends to 0, the spin angular momentum would also tend to zero. Orbital angular momentum survives in a classical world, because l can be allowed to approach infinity as $h \to 0$, and the quantity

$$\{l(l + 1)\}^{1/2} \frac{h}{2\pi} \approx l\frac{h}{2\pi}$$

can be non-zero. Uhlenbeck and Goudsmit's proposal was initially no more than a hypothesis, but when the English physicist Paul Dirac showed how to combine quantum mechanics and special relativity, the existence of particles with half-integral angular momentum quantum numbers appeared directly [1].

12.1.3 The width of an observed absorption

As in all measurements, there is an uncertainty associated with an observed absorption in atomic spectroscopy. And if absorption spectroscopy is to be used in establishing a frequency standard, and hence a time standard, the magnitude of the terms that contribute to the uncertainty associated with the absorption line centre must be estimated. The spectral line extends over a range of frequencies, not a single frequency; that is, it has a nonzero linewidth. In addition, its centre may be shifted from its nominal central wavelength. There are several reasons for this broadening and shift, which may be divided into two general categories: broadening due to local conditions and broadening due to extended conditions. Broadening due to local conditions is due to effects which exist in the neighbourhood of the emitting atom; usually small enough to assure local thermodynamic equilibrium. Broadening due to extended conditions may result from changes to the spectral distribution of the radiation as it traverses its path from the emitting atom to the detector.

The uncertainty principle relates the lifetime of an excited state (due to spontaneous radiative decay, which depends upon the transition dipole moment for the excitation) with the uncertainty of its energy. A short lifetime will have a large energy uncertainty, and a broad emission. This broadening effect results in an unshifted Lorentzian profile for the absorption. The natural broadening can be altered experimentally only to the extent that decay rates can be artificially suppressed or enhanced.

The atoms in a gas which are emitting radiation will have a distribution of velocities. Each photon emitted will be red- or blue-shifted by the Doppler effect depending on the velocity of the atom relative to the observer. The higher the temperature of the gas, the wider the distribution of velocities in the gas. Since the spectral line is a combination of all of the emitted radiation, the higher the temperature of the gas, the broader the spectral line emitted by that gas. This broadening effect is described by a Gaussian profile, and there is no associated frequency shift. Consequently, the most advantageous route to limit uncertainty in the measured line centre of the absorption is a measurement in a low-pressure gas at low temperature.

The presence of nearby particles will affect the radiation emitted by an individual particle. There are two limiting cases by which this occurs: via impact pressure broadening or collisional broadening. The collision of other particles with the emitting or absorbing atom interrupts the emission process and, by shortening the characteristic time for the process, increases the uncertainty in the energy emitted. Then there is quasistatic pressure broadening, where the presence of other particles shifts the energy levels in the emitting particle, via dispersion forces, thereby altering the frequency of the emitted radiation.

Certain types of broadening are the result of conditions over a large region of space rather than simply upon conditions that are local to the emitting particle. Electromagnetic radiation emitted at a particular point in space can be reabsorbed as it travels through space. This absorption depends on wavelength. The line is broadened because the photons at the line centre have a greater reabsorption probability than the photons at the line wings. Indeed, the reabsorption near the line centre may be so great as to cause a self-reversal in which the intensity at the centre of the line is less than in the wings. This process is also sometimes called self-absorption.

The details of line-shape in atomic spectroscopy can be found in any textbook of atomic spectroscopy [1].

12.1.4 Hydrogen maser

A hydrogen maser is also a frequency standard; it is actually an absorption in a hydrogen-atom. Indeed, the hydrogen maser is the most elaborate and readily available commercial frequency standard, or atomic clock. The word maser is an acronym that stands for **m**icrowave **a**mplification by **s**timulated **e**mission of **r**adiation, which defines the origin of the absorption seen at 1 420 405 752 Hz.

Consider figure 12.2. The ground state of a neutral hydrogen-atom consists of an electron bound to a proton. Both the electron and the proton have intrinsic magnetic dipole moments ascribed to their spins, whose interaction results in a slight increase in energy when the spins are parallel and a decrease when antiparallel. When the spins are parallel, the magnetic dipole moments are antiparallel (because the electron and proton have opposite electrical charge); thus, one would expect this configuration to actually have lower energy just as two magnets will align so that the north pole of one is closest to the south pole of the other. This model fails, because the wavefunctions of the electron and the proton overlap; that is, the electron is not spatially displaced from the proton, but encompasses it. The magnetic dipole moments are, therefore, best thought of as tiny current loops. As parallel currents attract, the parallel magnetic dipole moments (i.e. antiparallel spins) have lower energy. The transition between their orientational states described as a hyperfine transition has a frequency of 1420 MHz.

This transition is electric dipole forbidden with an extremely small transition rate of 2.9×10^{-15} s^{-1} and a mean lifetime of the excited state of around 10 million years (its decay is also electric dipole forbidden). Thus, a spontaneous occurrence of the transition is unlikely to be seen in a laboratory on Earth, but it can be induced using a hydrogen maser. It is, however, commonly observed in astronomical settings such as hydrogen clouds. Owing to its long lifetime, the line has an extremely small natural width, so most broadening is due to Doppler shifts caused by bulk motion, or nonzero temperature of the emitting regions.

In a hydrogen maser, a small storage bottle of molecular hydrogen (H_2) leaks a controlled amount of gas into a bulb. The molecules are dissociated in the bulb into individual hydrogen atoms via an electric discharge. This atomic hydrogen passes through a collimator and a magnetic state selector. The atoms are thereby selected for the desired state and passed on to a storage bulb, located inside a microwave

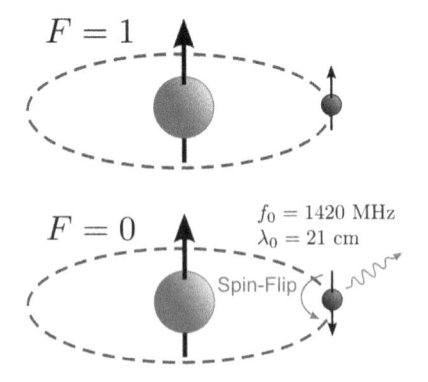

Figure 12.2. The origin of the 21 cm emission line from a hydrogen atom. The red sphere is the nucleus of the hydrogen atom, and the grey sphere is the electron of the Hydrogen atom. F is the quantum number defining the state of the total spin.

cavity. This cavity is tuned to the 1.420 GHz resonance frequency of the atoms. A weak static magnetic field is applied parallel to the cavity axis by a solenoid to lift the degeneracy of the magnetic Zeeman sublevels, permitting fine-tuning of the resonance.

12.1.5 Caesium beam oscillator

The hydrogen-atom has this hyperfine transition due to the interaction of the spins of its electron and proton. Similarly, a hydrogen-like atom such as caesium will also display this type of transition (in caesium, there are many more electrons, but these are all in closed shells except for a single lone s-orbital electron as in hydrogen). Caesium oscillators can also be primary frequency standards. Indeed, the present definition of the SI second is based upon the resonance frequency of the caesium atom (^{133}Cs), which is 9 192 631 770 Hz.

Inside a caesium oscillator, ^{133}Cs atoms are heated to a gaseous state in an oven (the metal boils at 670.8 °C). Atoms from the gas leave the oven in a high-velocity beam that travels through a vacuum toward a pair of magnets. The magnets serve as a selective gate that allows only atoms of a particular magnetic energy (Zeeman) state to pass through into a microwave cavity, where they are exposed to a microwave frequency derived from a quartz oscillator. If the microwave frequency matches the resonance frequency of caesium, the atoms change their magnetic state. Consider figure 12.3, the atomic beam then passes through a second magnetic state selector. Only those atoms that changed their magnetic state while passing through the intervening microwave cavity are allowed to proceed to a detector. Atoms that

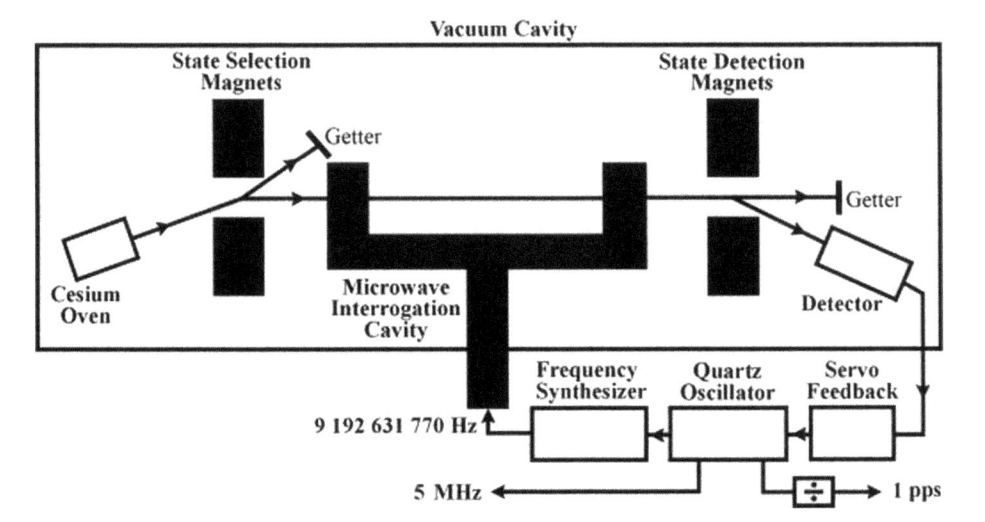

Figure 12.3. A design of an atomic beam magnetic resonance spectrometer. There are two magnetic state selectors and an intermediate region where the atoms undergo a change of state by application of a radiation field. (In the USA, caesium is spelt cesium.) (Image from https://www.nist.gov/pml/time-and-frequency-division/popular-links/time-frequency-z/time-and-frequency-z-c-ce; credit: National Institute of Standards and Technology.)

did not change state are deflected away from the detector. This is a classic Ramsey state selector [2]. The Q of a commercial caesium standard is a few parts in 10^8.[2] The beam tube is typically less than 0.5 m in length, and the atoms travel at velocities greater than 100 m s^{-1} in the vacuum tube. This limits the observation time to a few milliseconds, and the resonance width to a few hundred hertz. The stability at 1 s is typically 5×10^{-12}, and can reach a few parts in 10^{14} after one day of averaging. These are extraordinary levels of precision, when compared to other fields of metrology.

The resonance frequency of hydrogen is much lower than that of caesium, but the resonance width of a hydrogen maser is a few hertz. Therefore, the Q is about 10^9, or at least one order of magnitude better than that of a commercial caesium standard. As a result, the short-term stability is better than that of a caesium standard. However, when measured for more than a few days or weeks, a hydrogen maser might fall below a caesium oscillator's performance, due to changes in the cavity's resonance frequency over time.

12.1.6 Caesium fountain oscillator

The current state-of-the-art in caesium oscillator technology, the caesium fountain oscillator, is named for its fountain-like movement of caesium atoms. A caesium fountain, at the National Institute of Standards and Technology in the USA; NIST-F2 serves as the primary standard of time interval and frequency for the United States.

A caesium fountain works by sending a beam of caesium atoms into a vacuum chamber (by the adiabatic expansion of the atoms in a carrier gas as it expands through a small hole). As indicated in figure 12.4, six infrared laser beams are directed at right angles to each other at the centre of the chamber. By the force of absorption (the atoms move so slowly that they are perturbed by the momentum exchanged on absorbing a photon from the lasers), the lasers gently push the caesium atoms together into a ball—a form of condensation. In the process of creating this ball, the lasers slow down the movement of the atoms and cool them to temperatures a few millionths of a degree above absolute zero. This reduces their thermal velocity to a few centimetres per second (about 10^4 time slower than in a caesium beam oscillator).

A vertical laser beams gently 'tosses' the ball of condensed caesium atoms upward via an imparted forward momentum, and then all of the lasers are turned off. This little push is just enough to loft the ball about a metre, during which time it passes through a microwave cavity. Under the influence of gravity, the ball then stops and falls back down through the microwave cavity. The round trip, up and down

---

[2] Q factor is a parameter that describes the resonance behaviour of an underdamped harmonic oscillator (resonator). Sinusoidally driven resonators, having higher Q factors resonate with greater amplitudes (at the resonant frequency) but have a smaller range of frequencies around that frequency for which they resonate; the range of frequencies for which the oscillator resonates is called the bandwidth. Thus, a high-Q tuned circuit in a radio receiver would be more difficult to tune, but would have more selectivity; it would do a better job of filtering out signals from other stations that lie nearby on the spectrum.

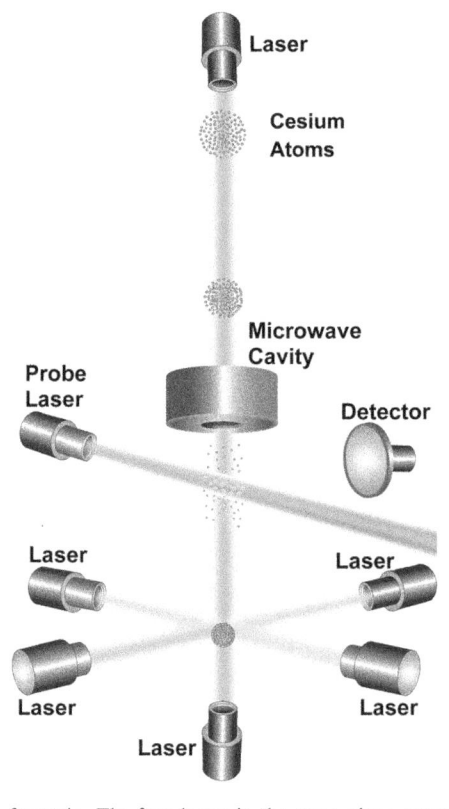

Figure 12.4. A caesium atom fountain. The four lasers in the same plane prepare the condensate of caesium atoms. This is lofted by the vertical leaser beam. The probe laser examines the quantum state of the caesium condensate after it has traversed the microwave radiation field. (Image from https://www.nist.gov/pml/time-and-frequency-division/popular-links/time-frequency-z/time-and-frequency-z-c-ce; credit: National Institute of Standards and Technology.)

through the microwave cavity lasts for about 1 s and is limited only by the force of gravity pulling the atoms downward. During the ascent and the descent, the magnetic (Zeeman) states of the atoms may or may not be altered as they interact with an applied microwave field. When the atomic excursion is finished, another laser investigates the quantum state of the atoms. Those atoms whose states were altered by absorption of the microwave radiation emit photons (they fluorescence) and are monitored by a detector. This process is repeated many times while the microwave signal in the cavity is tuned to different frequencies. Eventually, a microwave frequency is found that alters the states of most of the caesium atoms, and maximizes the observed fluorescence.

The Q of a caesium fountain is about 10^{10}, or about 100 times higher than a traditional, Ramsey caesium beam spectrometer (figure 12.3). Although the resonance frequency is the same, the resonance width is much narrower (<1 Hz), due to the longer observation times made possible by the combination of laser cooling and the fountain design. The combined frequency uncertainty of NIST-F2 is near 1×10^{-16}.

12.2 The future of frequency standards

Although caesium-fountain clocks are very accurate, there are limits to their precision. Collisions, for example, between cold caesium atoms in the fountain can shift the frequency of the atomic transition and broaden the observed linewidth. Second, the stabilities of one part in 10^{15} are only possible by averaging the signal over a period of about a day, which makes it hard to use the fountain clock, at this level of accuracy in real time. Optical clocks, however, could meet the need for better timekeeping. With frequencies approaching 10^{15} Hz, optical clocks should be stable to almost one part in 10^{15} simply by averaging over just a few seconds, rather than a day (in the optical region of the spectrum, there are more waves per centimetre). But with long averaging times, stabilities of one part in 10^{17}, or better should be possible. (See table 12.1 for a summary of the characteristics of these spectroscopic time-pieces.)

12.3 The mechanism of an optical clock

There are three main elements to an optical clock. The first is a highly stable reference frequency provided by a narrow optical absorption in an atom or ion. This 'clock transition' will typically have a natural line-width of a few hertz or less. The second element of the clock is a laser or local oscillator, which will have a line-width narrow enough so as not so to contribute to the natural line-width of the transition. The third component is related to the hands on a conventional clock-face and involves some way of counting the extremely rapid oscillations of the local oscillator; these oscillations are the ticks of the clock. A device known as a femtosecond comb, see figure 12.5, is used for this part of the device.

12.3.1 Femtosecond comb

A frequency comb is a laser source whose frequency spectrum consists of a series of discrete, equally spaced frequency lines; a femtosecond comb will have a frequency, or order, 10^{15} s^{-1}. Frequency combs can be generated by a number of mechanisms, including periodic modulation (in amplitude and/or phase) of a continuous-wave laser, four-wave mixing an nonlinear optical media, or stabilization of the pulse-train generated by a mode-locked laser [3].

The comb output displayed in figure 12.5 [4] is taken from a mode-locked femtosecond laser, which emits a train of pulses at a typical repetition rate, f_{rep}, of a few hundred megahertz. In the frequency domain, the sequence of pulses appears as a series of equally spaced frequencies. The frequency domain representation of a perfect frequency comb is a series of delta functions spaced according to

$$f_n = f_0 + n f_{rep},$$

where n is an integer and f_{rep} is the comb tooth spacing (equal to the mode-locked laser's repetition rate or, alternatively, the modulation frequency), and f_0 is the carrier offset frequency, which is less than f_{rep}. Combs spanning an octave in frequency (i.e. a factor of two) can be used to directly measure (and correct for drifts in) f_0. Octave-spanning combs can be used to steer a piezoelectric mirror within a

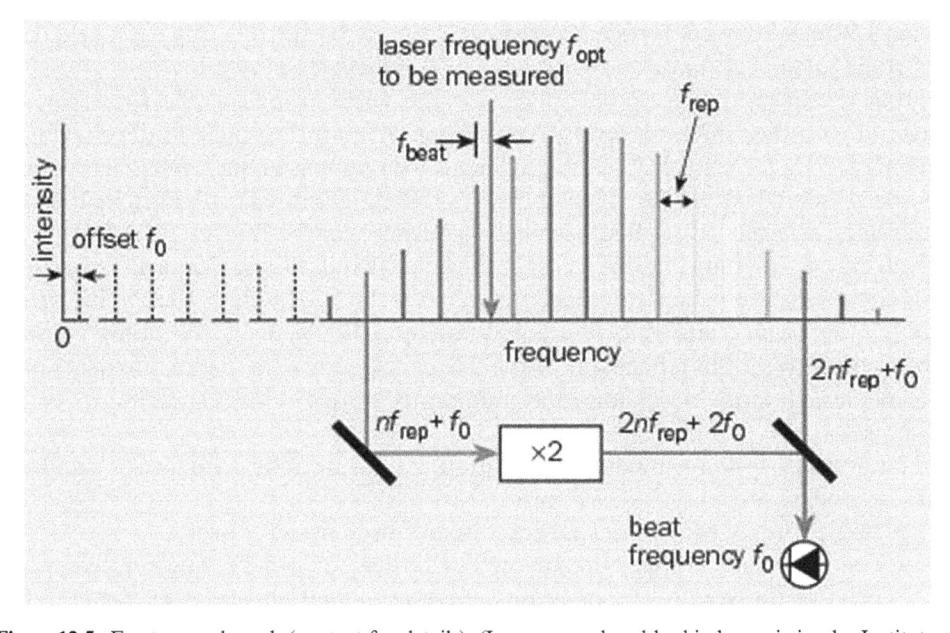

Figure 12.5. Femtosecond comb (see text for details). (Image reproduced by kind permission by Institute of Physics Publishing from *Physics World*, May 2005. Copyright 2005 IOP Publishing Ltd. Reproduced with permission. All rights reserved.)

carrier–envelope phase-correcting feedback loop. Any mechanism by which the combs' two degrees of freedom (f_0 and f_{rep}) are stabilized generates a comb that is useful for mapping optical frequencies into the radio frequency for the direct measurement of optical frequency [3, 4].

The frequency of any line in the comb is an integer multiple of the comb spacing (nf_{rep}) plus an offset frequency (f_0). The tick rate, f_{opt}, is related to f_{rep} and f_0, both of which can be determined experimentally. With f_{rep} and f_0 stabilized to an atomic clock, and thereby compared with the caesium primary frequency standard the comb can be used to measure the frequency of an optical standard f_{opt}. This is done by determining the beat frequency between the optical frequency and the precisely known frequency of the nearest comb mode. Alternatively, the comb may be stabilized to an optical standard rather than to a microwave standard. The comb then acts as the 'clockwork' of the optical clock. Critical to the performance of an optical clock is the first element mentioned above—the clock transition. This needs to be as narrow as possible to make the clock as possible. The ideal frequency reference would be a single, motionless atom or ion.

12.3.2 Optical clocks

The theoretical move from microwaves as the atomic 'escapement' for clocks to light in the optical range (harder to measure, but offering better performance characteristics– more waves per centimetre) earned John L Hall and Theodor W Hänsch the Nobel Prize in Physics in 2005 [3]. Then one of 2012's Physics Nobelists,

David J Wineland, went further by exploiting the properties of a single ion held in an electromagnetic trap to develop clocks of the highest stability [5].

Optical clocks are based on optical (in the visible range of the spectrum, see figure 11.1) rather than microwave transitions. A major obstacle to developing an optical clock is the difficulty of directly measuring optical frequencies. This problem has been solved with the development of self-referenced, mode-locked lasers; commonly referred to as femtosecond frequency combs (see above). Before the demonstration of the frequency comb, it was difficult to align optical frequencies with microwave frequency standards; terahertz techniques were needed to bridge the gap between radio and optical frequencies, and the systems for doing so were complicated. With the refinement of the frequency comb, these measurements have become much more accessible and numerous optical clock systems are being developed.

The primary systems under consideration for use in optical frequency standards (optical clocks) are:

- single multi-electron ions isolated in an ion Penning trap[3];
- neutral atoms trapped in an optical lattice formed by the interference of counter-propagating laser beams, creating a spatially periodic pattern. The resulting periodic potential may trap neutral atoms via the Stark shift, caused by the interaction of the atom's polarizability (α) and the oscillating electric field (E) of the light beams ($\approx \alpha E^2$). Atoms are cooled and congregate in the locations of potential minima. The resulting arrangement of trapped atoms resembles a crystal lattice;
- atoms packed in a three-dimensional quantum gas optical lattice.

These techniques allow the atoms or ions to be isolated from external perturbations, thus producing an extremely stable frequency reference; that is, the linewidth of the observed absorption is quantum limited, and not subject to, for example, broadening from collisions with other ions or atoms (see section 12.1.3).

Many multi-electron systems have been extensively studied as potential candidates for optical clocks: Al^+, $Hg^{+/2+}$, Hg, Sr, $Sr^{+/2+}$, $In^{+/3+}$, Mg, Ca, Ca^+, $Yb^{+/2+/3+}$, Yb, and $Th^{+/3+}$. The rare-earth element ytterbium (Yb) has been the object of extensive investigations [6]. An absolute frequency measurement of the unperturbed transition $^1S_0$–$^3P_0$ at 578 nm in ^{171}Yb was realized, in an optical lattice frequency standard relative to a cryogenic caesium fountain; the measured transition frequency is, 518 295 836 590 863.59(31) Hz with a relative standard uncertainty of 5.9×10^{-16} [6]. An equivalent measurement in strontium may also be found on the BIPM website [6].

---

[3] A Penning trap is a device for the storage of charged-particles, or a charged-particle using a homogeneous axial magnetic field and an inhomogeneous quadrupole electric field. This kind of trap is particularly well suited to precision measurements of properties of ions and stable subatomic particles. Single ions and single electrons have been isolated and studied this way to measure, for example, the electron magnetic moment. Such traps limit the time-broadening for absorption in the trapped single ions, that would otherwise reduce the precision of the measurement made in a bulk sample.

Figure 12.6. JILA's 2017 3-D quantum gas atomic clock consists of a grid of light formed by three pairs of laser beams. A stack of two tables is used to configure optical components around a vacuum chamber. Shown here is the upper table, where lenses and other optics are mounted. A blue laser beam excites a cube-shaped cloud of strontium atoms located behind the round window in the middle of the table. Strontium atoms fluoresce strongly when excited with blue light. (Image from https://en.wikipedia.org/wiki/ Atomic_clock#/media/File:17jila003_3d_strontium_atomic_clock.jpg; it has been obtained by the author from the Wikimedia website where it is stated to have been released into the public domain. It is included on that basis and is attributed to NIST.)

By 2013, optical lattice clocks were shown to be as good as, or even better than caesium fountain clocks [7]. Indeed, there was intense rivalry at this time to develop an optical clock that was so precise (of order, 4.3 parts in 10^{17} s) that the researchers could claim that such a clock would have lost less than a second over the age of the Universe (13.8 × 10^9 years or 4.3 × 10^{17} s). In 2015, JILA (a joint institute of the University of Colorado, Boulder, and NIST) evaluated the absolute frequency uncertainty of a strontium-87 optical lattice clock as 2.1 × 10^{-18}. In 2017, the JILA group reported [8] an experimental 3-D quantum gas strontium optical lattice clock in which strontium-87 atoms are packed into an ordered (by dispersion forces) 3-D cube at 1000 times the density of previous one-dimensional (1-D) clocks (figure 12.6). A synchronous clock comparison between two regions of the 3-D lattice yielded a record level of synchronization of 5 × 10^{-19} in 1 h of averaging time. The 3-D quantum gas strontium optical lattice clock's centrepiece is a state of matter called a degenerate Fermi gas (a quantum gas for Fermi particles; where all the atoms are in a single quantum level).

Optical clocks are currently still primarily research projects; the technology is less mature than caesium microwave standards, which regularly deliver time to the BIPM, for them to disseminate International Atomic Time. As the optical clocks move beyond their microwave counterparts in terms of accuracy and stability performance this puts them in a position to replace the current standard for time, the caesium fountain clock. In the future this might lead to redefinition of the SI second.

12.4 Secondary representations of the second

Since 2006, a list of frequencies recommended for secondary representations of the second is maintained by the BIPM, and is available, together with related documents at: https://www.bipm.org/en/publications/mises-en-pratique/standard-frequencies.html, see figure 11.1. The list contains the frequency values and the respective standard uncertainties for the rubidium microwave transition and for several optical transitions. These secondary frequency standards are accurate at the level of parts in 10^{-18}; however, the uncertainties provided in the list are in the range of parts in 10^{-14}–10^{-15} since they are limited by the linking to the caesium primary standard that currently defines the second. The metrological characteristics of such spectroscopic clocks is given in table 12.1.

Present-day optical atomic clocks that provide non-caesium-based secondary representations of the second have become so precise that they are likely to be used as extremely sensitive detectors for measurements other than determination of line positions; for example, the frequency of an optical atomic clock is altered slightly by gravity, magnetic fields, electrical fields, force, motion, temperature, and other phenomena. Given the levels of precision in our measurements, we are able to observe these minor perturbations on the transition frequency—having eliminated larger sources of uncertainty.

12.5 Possible applications of optical clocks

The development of microwave atomic clocks led to many scientific and technological advances; for example, a system of precise global and regional navigation satellites (the global positioning system or GPS), and applications via the Internet, which depend critically on frequency and time standards. Atomic clocks are installed at sites of time-signal radio transmitters. They are used at some long-wave and medium-wave broadcasting stations to deliver a very precise carrier frequency. Atomic clocks are used in many scientific disciplines, for example, for long-baseline interferometry in radio-astronomy. If the second were redefined with orders of magnitude greater precision, this added precision would be directly passed on to the present applications—present spatial resolution of the GPS system could be greatly improved.

Highly accurate optical clocks could also help to measure fundamental constants and test the laws of physics, such as Einstein's theories of special and general relativity. For example, some theorists believe that the fine-structure constant, α, which characterizes the strength of the electromagnetic interaction, may have changed over the history of the Universe [9]. If confirmed, the results would be of huge significance in cosmology and for theories that attempt to unify the four fundamental forces of Nature. One of the best ways of searching for variations of α is to compare the frequencies of different types of atomic clock over the course of several years. While this is a short period of time compared with cosmological timescales, this approach can provide results that are competitive with rival astrophysical measurements due to the extraordinarily high frequency-resolution achievable. The greater the precision of the clocks, the smaller the variations in α that could be detected, with the current limit approaching one part in 10^{15} per year.

Optical clocks could have other more practical applications; for example, while planes can navigate via GPS, it is not yet possible to land an aircraft by GPS alone, because the atomic clocks on satellites are still not accurate enough and it takes too long to compute positions—the spatial resolution is not good enough when it comes to 'touching down'. Highly accurate clocks will also be useful for deep-space probes, which need to travel vast distances. Moreover, improvements in the ground-based 'master clocks' that calibrate the GPS atomic clocks—along with better satellite clocks—will allow transportation systems to locate vehicles with sub-metre precision in real time.

Further reading

[1] Condon E U and Shortley G H 1977 *The Theory of Atomic Spectra* (Cambridge: Cambridge University Press)
   Carney A 1977 *Atomic and Laser Spectroscopy* (Oxford: Oxford University Press)
[2] Ramsey N F 1969 *Molecular Beams* (Oxford: Oxford University Press)
[3] Hall J L 2006 Nobel Lecture: Defining and measuring optical frequencies *Rev. Mod. Phys.* **78** 1279–95
   Hänsch T W 2006 Nobel Lecture: Passion for precision *Rev. Mod. Phys.* **78** 1297–309
[4] Gill P and Margolis H 2005 Optical clocks *Phys. World* https://physicsworld.com/a/optical-clocks/
[5] Wineland D J Nobel Lecture, *Superposition, Entanglement, and Raising Schrödinger's Cat*; https://nobelprize.org/uploads/2018/06/wineland-lecture.pdf
[6] See the BIPM's list of recommended radiations for the practical realization of the metre and the secondary representation of the definition of the second. The entry for the $6s^2$ ^{1}S0—$6s6p$ ^{3}P0 transition in ^{171}Yb is https://bipm.org/utils/common/pdf/mep/171Yb_518THz_2013.pdf and for ^{87}Sr is https://bipm.org/utils/common/pdf/mep/87Sr_429THz_2015.pdf
   Campbell G K *et al* 2008 The absolute frequency of the ^{87}Sr optical clock transition *Metrologia* **45** 539–48
[7] Bloom B J, Nicholson T L, Williams J R, Campbell S L, Bishof M, Zhang X, Zhang W, Bromley S L and Ye J 2014 A new generation of atomic clocks: Total uncertainty and instability at the 10^{18} level *Nature* **506** 71–5
[8] Campbell S L, Hutson R B, Marti G E, Goban A, Oppong N D, McNally R L, Sonderhouse L, Zhang W, Bloom B J and Ye J 2017 A Fermi-degenerate three-dimensional optical lattice clock *Science* **358** 90–4
[9] Webb J 2003 Are the laws of nature changing with time? *Phys. World* 33–8 https://physicsworld.com/a/are-the-laws-of-nature-changing-with-time/

Chapter 13

The new Système international des unites

The old SI is dead—long live the Quantum-SI! The previous two chapters have outlined how the international science community has moved away from defining the base units of the SI in terms of artefacts and near-artefact experimental procedures. Today, we have a new SI, a Quantum-SI defined in terms of constants of Nature. In this chapter, we will look in more detail at these changes, and briefly consider how the various constants of Nature are measured.

13.1 Some further details of the Quantum-SI

20 May 2019 was not only the 144th anniversary of the signing of the Metre Convention in Paris, intended to promote the use of the Metric System throughout the world, it also marked the biggest change to the modern scientific representation of the Metric System, the International System of Units (SI) for well over half a century. The history of the SI can be seen in the bullet points towards the end of chapter 8. The previous major change of the scientific aspect of the Metric System occurred in 1960 when the SI was formally created and given a coherence. At that time, the metre was redefined, abolishing the International Prototype of the Metre, which had been kept at the BIPM and promulgating a definition of the metre based on a certain number of wavelengths of the red-light emitted by a krypton-86 atomic discharge lamp. Thereby permitting the realisation of the metre in any laboratory, and removing the primacy of the artefact in Sèvres. In 1960, the kilogram, however, remained defined by a physical prototype.

The original SI of 1960 was a coherent system constructed around seven base units, powers and multiples of which were used to construct all other units. With the redefinitions of 2019, the SI is now constructed around seven defining natural constants (see figure 8.1). Permitting all units to be constructed directly from these constants. The designation of base units is retained, but is no longer essential to define SI measurements. As a consequence of the recent redefinition, four of the seven SI base units—the kilogram, ampere, kelvin, and mole—were redefined by setting exact numerical values for the Planck constant (h), the elementary electric charge (e), the Boltzmann constant (k_B), and the Avogadro constant (N_A), respectively. The second, the metre, and the candela were already defined by physical constants and their definitions were subject to correction in 2019. The new definitions seek to improve the SI without changing the value of any units, thereby ensuring continuity with existing measurements. This process of redefinition followed the route of any important international decision concerning metrology; that is, it was instigated and organized by the institutions created under the Metre Convention of 1875 (see chapter 7). In November 2018, the 26th General Conference on Weights and Measures (CGPM) unanimously approved the proposed redefinitions changes, which the International Committee for Weights and Measures (CIPM) had proposed earlier that year after determining that previously agreed conditions for the change had been met. These conditions were satisfied by a series of experiments that measured the required constants of Nature to a sufficiently high level of precision relative to the old SI definitions. This was the culmination of decades of research, and the CGPM decreed the new definitions would enter into force; that is, become mandatory on 20 May 2019. [1]

The Metric System was originally conceived as a system of measurement that derived from unchanging phenomena, but practical limitations necessitated the use of artefacts—the prototype metre and prototype kilogram—when the Metric System was introduced in France in April 1795. Although it was designed for long-term stability, the masses of the prototype kilogram and its secondary copies have shown small variations, relative to each other over time. This uncertainty in the definition of mass was considered to be unacceptable for use in the wider scientific community, and a more stable definition of mass was sought. This has now been achieved, and in various sections of this present work, one may find details about all the redefinitions, and the creation of a new SI.

It must not be thought, however, that little had occurred between the creation of the SI in 1960, and the creation of the Quantum-SI in 2019. The SI has undergone a great many changes. The process of change began with the 1983 redefinition of the metre in terms of an exact numerical value for the speed of the red-light emitted by a stabilised He–Ne laser. The BIPM's Consultative Committee for Units (CCU) went on to recommend, and the BIPM proposed that four further constants of Nature should be defined to have exact values. These were:

- The Planck constant, h, is exactly $6.626\ 070\ 15 \times 10^{-34}$ joule second (J s).
- The elementary charge, e, is exactly $1.602\ 176\ 634 \times 10^{-19}$ coulomb (C).
- The Boltzmann constant, k_B, is exactly $1.380\ 649 \times 10^{-23}$ joule per kelvin (J K^{-1}).
- The Avogadro constant, N_A, is exactly $6.022\ 140\ 76 \times 10^{23}$ reciprocal mole (mol^{-1}).

These constants are described in the 2006 version (8th edition) of the *SI Brochure* [2] but in that edition the latter three quantities are defined as *constants to be obtained by experiment* rather than as *defining constants*. The 2019 redefinition, and the 9th edition of the *SI Brochure*, retains the numerical values associated with the following constants of Nature [3]:

- The speed of light, c, is exactly 299 792 458 m per second (m s^{-1}).
- The ground state hyperfine structure transition frequency of the caesium-133 atom, $\Delta\nu_{Cs}$, is exactly 919 263 177 0 hertz (Hz).
- The luminous efficacy, K_{cd}, of monochromatic radiation of frequency 540×10^{12} Hz, a frequency of green light at approximately the peak sensitivity of the human eye is exactly 683 lumens per watt (lm W^{-1}).

These seven definitions are rewritten below with the derived units (joule, coulomb, hertz, lumen, and watt) expressed in terms of the seven base units; second, metre, kilogram, ampere, kelvin, mole, and candela, according to the 9th edition of the *SI Brochure* [3]. In the following list, the symbol sr stands for the dimensionless unit steradian. Note also the very different levels of precision associated with these, now fixed constants; the frequency of the hyperfine transition in a ^{133}Cs atom is quoted to ten significant figures, but the luminous efficacy, K_{cd}, may only be quoted to three significant figures.

- $\Delta\nu_{Cs} = \Delta\nu(^{133}\text{Cs})_{hfs} = 919\ 263\ 177\ 0\ \text{s}^{-1}$
- $c = 299\ 792\ 458\ \text{m s}^{-1}$
- $h = 6.626\ 070\ 15 \times 10^{-34}\ \text{kg m}^2\ \text{s}^{-1}$
- $e = 1.602\ 176\ 634 \times 10^{-19}\ \text{A s}$
- $k = 1.380\ 649 \times 10^{-23}\ \text{kg m}^2\ \text{K}^{-1}\ \text{s}^{-2}$
- $N_A = 6.022\ 140\ 76 \times 10^{23}\ \text{mol}^{-1}$
- $K_{cd} = 683\ \text{cd sr s}^3\ \text{kg}^{-1}\ \text{m}^{-2}$

As part of the redefinition, the International Prototype of the Kilogram (see figure 7.4) was retired, and the existing definitions of the kilogram, the ampere, and the kelvin were replaced. The definition of the mole was revised; see table 11.1 for full details of the redefinitions. These changes have the effect of redefining the SI base units, although the definitions of the SI derived units in terms of the base units remain the same. The values of these constants are those recommended by the 2018 CODATA (see footnote in section 11.4) [4–7].

As the recent redefinitions of the quantities of the SI have fixed the values of certain constants of Nature, let us briefly look at the present means of measuring some of these constants.

13.2 Experimental measurements of the elementary charge, *e*

13.2.1 In terms of the Avogadro constant and Faraday constant

If the Avogadro constant N_A and the Faraday constant F are known independently, the value of the elementary charge, e may be deduced from: $e = F/N_A$, that is, the charge of one mole of electrons, divided by the number of electrons in a mole equals the charge of a single electron. This method does not yield the most precise values of the constant on interest, but it is an accurate method that permits one to relate properties of bulk samples to the properties of the individual molecules comprising the sample [8].

The value of the Avogadro constant N_A was first approximated by the German chemist Johann Josef Loschmidt (1821–95) who, in 1865, estimated the average diameter of the molecules in air by a method that is equivalent to calculating the number of particles in a given volume of gas [8]. Today the value of N_A can be measured to high accuracy by taking an extremely pure crystal (isotopically pure silicon) and measuring the volume of the unit cell of the crystal (see section 13.5), but one must know the density of the crystal. From this information, one can deduce the mass (m) of a single atom; and since the molar mass (M) is known, the number of atoms in a mole may then be calculated, $N_A = M/m$.

The research of Loschmidt is worth a slight detour as it is fundamental to the acceptance of the reality of molecules by the chemistry community in the latter-half of the 19th Century, and by the physics community in the years after World War I. This research also tells us something about using experiment to determine unknown quantities. In 1866, the Loschmidt calculated the diameter of an 'air molecule' and concluded that 'in the domain of atoms and molecules, the appropriate measure of length is the millionth of a millimetre'. Four years later, William Thomson (later Lord Kelvin) who regarded it 'as an established fact of science that a gas consists of moving molecules', also measured the diameters of gaseous molecules. The experimental scientists of the day were seeking to measure the increasing small; and in so doing they made significant discoveries [8].

Of particular importance were the investigations of Loschmidt, who demonstrated that if one had two unknown quantities to be determined, one needed more than one experimental observation. This may seem obvious today, but it was less obvious back in 1866. Loschmidt wished to determine both the diameter of a molecule and a value for Avogadro's number (N_A). He had an expression which related these two unknowns to an experimental observation; however, he found that he required two simultaneous equations in which the two unknowns (N_A and the molecular diameter) are expressed in terms of known physical quantities. The first equation used by Loschmidt to determine his two unknowns is the relation between the molecular diameter (d), the mean free path (λ), and the number (n) of molecules or atoms per unit volume of an ideal gas: $\lambda n d^2 \pi = $ a calculable constant. His second equation concerned the fraction of space occupied by a hard-sphere, or ideal gas, $n\pi d^3$. Thus, Loschmidt had two equations for n (and hence N_A) and d. At the end of the 19th Century, Albert Einstein was to use a similar rational in rationalising Brownian motion [8].

The value of F can also be measured directly using Faraday's laws of electrolysis, which provide a quantitative basis to the electrochemical experiments of Michael Faraday in 1834. In an electrolysis experiment, there is a one-to-one correspondence between the electrons passing through the anode-to-cathode wire and the ions that plate onto, or off of the anode or cathode. Measuring the change in mass of the anode or cathode, and the total charge passing through the wire (which can be measured as the time-integral of electric current), and also taking into account the molar mass of the ions, one may deduce F.

13.2.2 Oil-drop experiment

A famous method for measuring e is the oil-drop experiment of Robert Millikan (1868–1953); originally undertaken in 1909. A falling small-drop of oil subjected to an electric field would move at a rate that balanced the forces of gravity, viscosity (of traveling through the air), and the electric force. The forces due to gravity and viscosity may be calculated from the size and velocity of the oil drop, so the electric force may be deduced. Since electric force is the product of any residual electric charge on the droplet, and the known applied electric field the electric charge of the droplet can be computed. By measuring the charges of many different oil drops, it can be seen that the charges are all integer multiples of a single charge, namely e. The necessity of measuring the size of the oil droplets can be eliminated by using tiny plastic spheres of a uniform size. The force due to viscosity can be eliminated by adjusting the strength of the electric field so that the sphere hovers motionless in the applied field.

13.2.3 Shot noise

The magnitude of an electric current in an electronic device (or even a wire) will be associated with noise from a variety of sources; one of which is shot noise. Shot noise exists because a current is not a smooth continual flow; instead, a current is made up of discrete electrons that pass by one at a time. By carefully analysing the noise of a current in a circuit, the charge of an electron can be calculated. This method, first proposed by Walter H Schottky, can determine a value of e of which the accuracy is limited to a few percent in a non-quantum environment.

13.2.4 The Josephson and von Klitzing constants

Another accurate method for measuring the elementary charge is by inferring it from measurements of two quantum phenomena: The Josephson effect; voltage oscillations that arise in certain superconducting structures, and the quantum Hall effect; a quantum effect of electrons at low temperatures, strong magnetic fields, and confinement into two dimensions (see chapter 11). The Josephson constant is

$$K_J = 2e/h,$$

where h is the Planck constant. Knowing h, one can measure e directly. The von Klitzing constant is

$$R_K = h/e^2.$$

From these two quantum constants, the elementary charge can be deduced[1]:

$$e = 2/K_J R_K$$

---

[1] The simple relationships of K_J and R_K, with respect to e and h are generally held to be exact. However, they are both based on superconducting environments where we are not speaking of the properties of electrons, but of Cooper pairs. Cooper pairs are not elementary particles, but exist only through the coupling of the two electrons involved with the distortions of the surrounding lattice. So, the binding energy between the electrons in the Cooper pair contains terms involving polarization and induction; that is, strong perturbations of the containing lattice.

13.3 The problem of the permeability of space in the new SI

In SI units, permeability is measured in henries per metre (H m^{-1}), or equivalently in newtons per ampere squared (N A^{-2}). The permeability constant μ_0, also known as the magnetic constant, or the permeability of free-space is a measure of the amount of resistance encountered when forming a magnetic field in a classical vacuum.

The magnetic constant μ_0 appears in Maxwell's equations in quantities such as permeability and magnetization density (see chapter 6); for example, the relationship that defines the magnetic H-field in terms of the magnetic B-field. In real media, this relationship has the form

$$H = \left(B/\mu_0\right) - M, \tag{13.1}$$

where M is the magnetization density; in vacuum, $M = 0$.

In SI units, the speed of light in vacuum, c, is related to the magnetic constant and the electric constant (vacuum permittivity), ε_0, by the definition

$$c = 1/\sqrt{\mu_0 \varepsilon_0}. \tag{13.2}$$

This relation can be derived via Maxwell's equations of classical electromagnetism in the classical vacuum, but this relation is used as a definition of ε_0 in terms of the defined numerical values for c and μ_0, and is not presented as a derived result contingent upon the validity of Maxwell's equations. Conversely, as the permittivity is related to the fine structure constant (α), the permeability can be derived from the latter, using the Planck constant, h, and the charge of the electron, e [9]

$$\mu_0 = 2\alpha h/e^2 c. \tag{13.3}$$

Before 20 May 2019, the magnetic constant had the exact defined value $\mu_0 = 4\pi \times 10^{-7}$ H m$^{-1} \approx 12.57 \times 10^{-7}$ H m^{-1}. The old definition of the ampere (see table 11.1) was such that it defined an exact value for this constant, as a force between two current-carrying wires a metre apart. Since May 20, 2019, however, the vacuum permeability μ_0 is no longer a defined constant, as *per* the old definition, but rather a quantity that will need to be determined experimentally. The 2018 CODATA value for μ_0 is given below; this quantity is proportional to the dimensionless fine-structure constant with no other dependences

$$\mu_0 = 1.256\ 637\ 062\ 12(19) \times 10^{-6}\ \text{H m}^{-1}.$$

In the old-SI, in the reference medium of the classical vacuum, μ_0 had an exact defined value: $\mu_0 = 4\pi \times 10^{-7}$ H m$^{-1} = 1.256\ 637\ 0614\ \ldots \times 10^{-6}$ N A^{-2} [9]. So what was a fixed constant, by the old definition of the ampere is now a measurable quantity, possessing an associated uncertainty.

13.4 Determination of the Planck constant

The Planck constant, h has dimensions of physical action; i.e., energy multiplied by time, or momentum multiplied by distance; that is, angular momentum. In SI units, the Planck constant is expressed in joule seconds (J s, or N m s, or kg m^2 s^{-1}). Implicit in the dimensions of the Planck constant is the fact that the SI unit of frequency, the Hertz, represents one complete cycle, 360 degrees or 2π radians per second. An angular frequency in radians per second is often more natural in mathematics and physics and many formulae use a reduced Planck constant ($\hbar = h/2\pi$).

In principle, the Planck constant can be determined by examining any of the phenomena explicable through quantum mechanics; for example, the spectrum of a black-body radiator, or the kinetic energy of photoelectrons [10]. Indeed, this is how its value was first determined in the early-years of the last century; however, these are no longer the most accurate methods. Max Planck was investigating the energy content of the radiation emitted by a black-body, and came to the conclusion that there could be no equilibrium between matter and radiation unless we suppose that the interaction between matter and radiation takes place, not as had been thought continuously, but as a series of steps. A small amount (a quantum) of energy being transferred from matter to radiation, or vice versa in interactive steps. [8] Since the value of the Planck constant is now fixed by the definition of the kilogram, it is no longer measured. This will remain the case until such time as new paradigms are derived, or a more complete synthesis of the interaction of the constants of Nature achieved. Intriguingly, many of the experimental procedures given here to previously determine a value for the Planck constant may now be used to determine the mass of the kilogram (see chapter 11).

13.4.1 Josephson constant

The Josephson constant K_J relates the potential difference U generated by the Josephson effect, at a Josephson junction with the frequency ν of the microwave radiation used to excite the junction. The theoretical treatment of Josephson effect suggests very strongly that $K_J = 2e/h$; e being the electronic charge (see footnote in section 13.2.4). The Josephson constant may be measured by comparing the potential difference generated by an array of Josephson junctions with another potential difference. The measurement of the potential difference is made by allowing an electrostatic force to cancel out a measurable gravitational force in a Kibble balance. Assuming the validity of the theoretical treatment of the Josephson effect, K_J is related to the Planck constant by

$$h = 8\alpha/\mu_0 c K_J{}^2,$$

where α is the fine-structure constant, also known as Sommerfeld's constant; it is a fundamental physical constant characterizing the strength of the electromagnetic interaction between elementary charged particles. It is a dimensionless quantity related to the elementary charge e, which characterizes the strength of the coupling of an elementary charged-particle with the electromagnetic field, by the formula

$4\pi\varepsilon_0\hbar c\alpha = e^2$. As a dimensionless quantity, it has the same numerical value whatever system of units is being used, and is nearly $1/137^2$.

13.4.2 Kibble balance

A Kibble balance (formerly known as a watt balance) is an instrument for comparing two quantities of power; one of which is measured in SI watts and the other of which is measured in conventional electrical units. From the definition of the conventional watt W_{90}, this gives a measure of the product $K_J^2 R_K$ in SI units, where R_K is the von Klitzing constant, which appears in the quantum Hall effect. If the theoretical treatments of the Josephson effect and the quantum Hall effect are valid, and in particular assuming that $R_K = h/e^2$, the measurement of $K_J^2 R_K$ is a direct determination of the Planck constant, $h = 4/ K_J^2 R_K$ [7]. A full description of the Kibble balance is given in chapter 11.

13.5 Measurement of N_A by x-ray diffraction

Today, it is through x-ray diffraction measurements, undertaken on isotopically pure crystals of silicon that the most precise and reliable values of Avogadro's number are obtained[3]. Such diffraction experiments measure the volume of the unit cell of the silicon crystal. The crystal is cubic, so the unit cell may be defined by one length measurement. A single crystal of silicon may be produced with extremely high isotopic purity, and few lattice defects. This diffraction method defines the Avogadro constant as the ratio of the molar volume, V_m, to the atomic volume V_{atom}: $N_A = V_m/V_{atom}$, where $V_{atom} = V_{unit\ cell}/n$ and n is the number of atoms per unit cell of volume V_{cell}.

The unit cell of silicon has a cubic packing arrangement of 8 atoms, and the unit cell volume may be measured by determining a single unit cell parameter, a (see figure 13.1). In practice, measurements are carried out on a distance known as, $d_{220}(Si)$, which is the direction between the planes denoted by the Miller indices {220} and is equal to $a/\sqrt{8}$. The 2006 CODATA value (see footnote in section 11.4.1) for $d_{220}(Si)$ is 192.015 576 2(50) pm, a relative standard uncertainty of 2.8×10^{-8}, corresponding to a unit cell volume of $1.601\ 933\ 04(13) \times 10^{-28}\ m^3$.

The isotope composition of the sample used must also be measured and taken into account in the analysis of the diffraction experiment. Silicon occurs in three

---

[2] For many years, it appeared that the quantity $1/\alpha$ might be an exact integer, 137. This number, caused speculation [11], sleepless nights, mystic visions and one of the best jokes ever to find its way through the peer-review system [12]. Hans Bethe was well-known for his sense of humour, and with Guido Beck, Wolfgang Riezler, and two postdocs he created a hoax paper *On the Quantum Theory of the Temperature of Absolute Zero*, where he calculated the fine structure constant (α) from the absolute zero temperature in Celsius units. The paper poked fun at a certain class of papers in the theoretical physics of the day, which were purely speculative, and based on spurious numerical arguments such as Arthur Eddington's attempts to explain the value of the fine structure constant from fundamental quantities. Bethe and his co-authors were forced to issue an apology. All this subsided when it was found, by experiment that $(1/\alpha) = 137.035\ 999\ 084(21)$. As a dimensionless quantity, it has the same numerical value in whichever system of units is used.

[3] The references in [8] provide a history of the evolution of our acceptance of the existence of molecules, which is intimately related to the evolution of the study of Avogadro's number—the molecule meme.

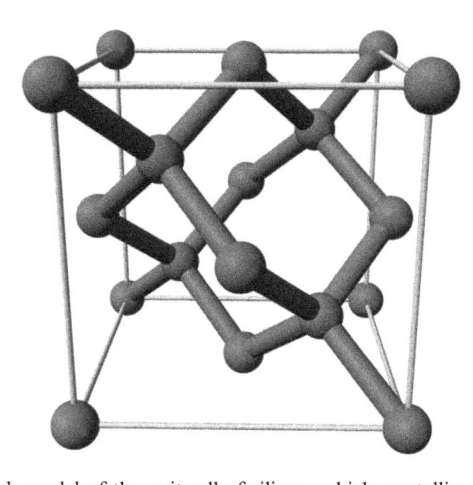

Figure 13.1. A ball-and-stick model of the unit cell of silicon, which crystallizes in a diamond cubic lattice, with one unit cell dimension; that is, $a = b = c = 5.4309$ Å.

stable isotopes (^{28}Si, ^{29}Si, and ^{30}Si), and the natural variation in their proportions is greater than other uncertainties in the measurements. The atomic weight A_r for the sample crystal can be calculated, as the standard atomic weights of the three nuclides are known with sufficient precision. This, together with the measured density ρ of the sample, allows the molar volume V_m to be determined $V_m = A_r M_u / \rho$. From the 2006 CODATA compilation of the best experimental data, the relative uncertainty in determinations of the Avogadro constant by the x-ray crystal density method is 1.2×10^{-7}.

The International Avogadro Coordination, sometimes called the Avogadro project, is a collaboration begun in the early-1990s between various national metrology institutes to measure the Avogadro constant by the x-ray crystal density method to a relative uncertainty of 2×10^{-8} or less. The project is part of the effort to redefine the SI-kilogram in terms of a physical constant, rather than the artefact known as the International Prototype Kilogram, which is kept in a safe in Sèvres near Paris, and complements the measurements of the Planck constant using watt balances. Under the current definitions of the SI, a measurement of the Avogadro constant is an indirect measurement of the Planck constant $h = c\alpha^2 A_r(e) M_u / 2 R_\infty N_A$, where R_∞ is the Rydberg constant, α is the fine-structure constant, and $A_r(e)$ is the 'relative atomic mass' of an electron.

The diffraction measurements to determine N_A use highly polished spheres of silicon with a mass of 1 kg. Spheres are used to simplify the measurement of the size (and hence the density) and to minimize the effect of the oxide coating that inevitably forms on the surface of such highly polished objects. The first measurements used spheres of silicon with natural isotopic composition and had a relative uncertainty of 3.1×10^{-7}. These first results were also inconsistent with values of the Planck constant derived from watt balance measurements, although the source of the discrepancy was identified.

The main residual uncertainty in the early x-ray measurements of N_A was in the measurement of the isotopic composition of the silicon to calculate the atomic weight. Consequently, in 2007 a 4.8 kg single crystal of isotopically enriched silicon (99.94% ^{28}Si) was grown, and two one-kilogram spheres cut from it (see figures 14.1 and 14.2). Diameter measurements on the spheres are repeatable to within 0.3 nm, and the uncertainty in the mass is 3 μg (3×10^{-6}). The details mentioned above describe the first generation of silicon spheres; for the recent redefinition of the SI, a new generation of silicon spheres was fabricated [13].

The present definition of the mole determines the value of the constant of Nature that relates the number of entities to amount of substance for any material. This is Avogadro's constant, N_A. If $N(X)$ denotes the number of entities X in a specified sample, and if $n(X)$ denotes the amount of substance of entities X in the same sample, we may write $n(X) = N(X)/N_A$. Note that since $N(X)$ is dimensionless, and n (X) has the SI unit, the mole, the Avogadro constant has the coherent SI unit reciprocal mole. This relationship is the basis of all chemistry, and one of the quantities that bring chemistry and physics together.

Further reading

[1] Stock M, Davis R, de Mirandés E and Milton M J T 2019 The revision of the SI—the result of three decades of progress in metrology *Metrologia* **56** 022001

[2] *The International System of Units* (*SI*), 2006, 8th ed. Published by the BIPM and available from the BIPM

[3] *The International System of Units* (*SI*), 2019, 9th ed. Published by the BIPM and freely available on their website https://www.bipm.org/en/publications/si-brochure/

[4] Mohr P J, Newell D B and Taylor B N 2016 CODATA 2014 recommended values of the fundamental constants *Rev. Mod. Phys.* **88** 035009

[5] Newell D B *et al* 2018 The CODATA 2017 values of h, e, k_B, and N_A for the revision of the SI *Metrologia* **55** L13

[6] Mohr P J, Newell D B, Taylor B N and Tiesinga E 2018 Data and analysis for the CODATA 2017 special fundamental constants adjustment *Metrologia* **55** 125

[7] Possolo A, Schlamminger S, Stoudt S, Pratt J R and Williams C J 2018 Evaluation of the accuracy, consistency, and stability of measurements of the Planck constant used in the redefinition of the international system of units *Metrologia* **55** 29

[8] Williams J H 2018 *Molecule as Meme* (San Jose, CA: Morgan & Claypool);
Pais A 1982 *Subtle is the Lord: The Science and the Life of Albert Einstein* (Oxford: Oxford University Press) chapter 5, Reality of molecules

[9] Davis R S 2017 Determining the value of the fine-structure constant from a current balance: Getting acquainted with some upcoming changes to the SI *Am. J. Phys.* **85** 364–8

[10] Pais A 1982 *Subtle is the Lord: The Science and the Life of Albert Einstein* (Oxford: Oxford University Press)

[11] Eddington A S 1936 *Relativity Theory of Protons and Electrons* (New York: Macmillan)

[12] Corlin A, Stein J S, Beck G, Bethe H A and Riezler W 1931 *Naturwissenschaften* **19** 37–9

[13] Abrosimov N V *et al* 2017 A new generation of 99.999% enriched ^{28}Si single crystals for the determination of Avogadro's constant *Metrologia* **54**

Chapter 14

For this is science

14.1 Units of measurement must evolve, because science evolves

Although the Taoists of Ancient China and the natural philosophers of the classical world were the first to observe Nature, so as to seek reassurance against their fear of the power of natural phenomena, it was the *savants* and, particularly, the experimental natural philosophers of the 17th Century who began investigating and applying the newly discovered laws of Nature. These *savants* and natural philosophers of the early modern period displaced the magicians, shamans and astrologers, whose predictions of future events were correct only on a statistical basis.

The essential character of science is that it is predictive. It follows some established order; thought originally to be divinely inspired, and which is still something that we have yet to fully comprehend. We hold that phenomena arise because of a set of fundamental laws, and the interaction between a set of unchanging constants of Nature. We further hold that the laws of physics are inviolable. And that these laws and their associated constants convey an absolute authority. Whether we know it or not, and irrespective of whether we like it or not, these laws and constants of Nature shape our lives. The laws of physics, and the constants of Nature define our very morphology. The precise distances and orientations between the molecules of which our bodies are composed are determined by subtle intermolecular electrostatic forces, whose magnitude is determined by the various constants of Nature; particularly, the mass and charge of the electron, and whose function is dictated by the laws of physics. We are merely living, tangible representations of those physical laws.

It was, however, only with the success of the scientific rationalism of the 19th Century that we moved definitively to speaking about the laws of Nature, and the advent of science as a language of authority capable of explaining the world around us. It could not be otherwise, we knew what was happening everywhere in Creation, because it happened in our laboratories here on Earth. To have spoken of rules or propositions of Nature would have been humbler; we wished, however, to say that science (and by implication, the scientist) was the new omnipotent force.

Science is capable of directly answering many of the questions which we pose when we stare out into space and consider our own place in the Cosmos, but there are no absolute truths, and one must always be aware that even in science there is doubt and uncertainty; the scientific world-view is continually evolving. Scepticism and questioning are essential qualities in a scientist. At the end of the 19th Century, Max Planck attempted to measure the energy content of long-wavelength radiation emitted by a hot-body. In fact, what he was doing was questioning the nature of measurement itself. If you wish to measure the temperature of a body, how small a perturbation of that body can you make and still obtain the measurement you desire, without having changed the body you are studying to such an extent that your measurement becomes useless? What is the size and nature of this tiny amount of energy that you can remove from a body without changing the body irredeemably?

This research on the measurement of tiny fluctuations of temperature and energy (both of which are related through the constant of Nature known as Boltzmann's constant) led to the development of new technologies for making such measurements. This in turn led to the desire to make increasingly precise measurements, which triggered the development of even more precise techniques of measurement, and so on. The scientists' questions were driving the development of technology by always seeking to push the measurement horizon. This research on temperature showed us in detail how energy flows through matter and led Max Planck to the concept of the *quantum* or the smallest possible quantity of energy, and then to quantum mechanics; arguably the most successful theory ever devised in the physical sciences.

Another such pioneer of measurement was Samuel Pierpont Langley (1834–1906); an American astronomer, a founder of aviation and Secretary of the Smithsonian Institute where he invented the bolometer; an instrument devised to measure the energy of infrared radiation.

Today, the bolometer is a well-accepted tool for fundamental investigations in chemical physics and astronomy. However, when the first bolometer was built in Washington City in the late-19th Century, Langley's contemporaries (particularly his friend, the historian Henry Adams) ridiculed him as 'the man attempting to measure nothing'; which was exactly what he was trying to do. Langley was pushing the technology for the measurement of minute changes in temperature beyond its then limits; he was indeed attempting to measure something close to nothing.

For all the joking, Langley made the first measurement of the surface temperature of the Moon by looking at the effect of moonlight (which is principally infra-red radiation associated with the glorious vision we see in the visible range of the electromagnetic spectrum) in his bolometer. One can readily imagine the mirth with which his experiment on moonshine must have been greeted by his sophisticated, non-physicist friends.

Langley showed that the temperature of the Lunar surface measured from the intensity of moonlight entering his bolometer varied depending upon the angle of incidence of the Moonlight relative to his bolometer; that is, the depth of the Earth's atmosphere that had been traversed by the infrared radiation coming from the Moon (the pathlength of the radiation). Langley went on to show that this angular variation was due to the amount of carbon dioxide in the Earth's atmosphere. In addition, Langley's measurements of the interference of the Sun's infrared radiation incident upon the Earth's surface by carbon dioxide in the Earth's atmosphere was used by the Swedish chemist Svante August Arrhenius (1859–1927) in 1896 to make the first calculation of how the world's climate would change from a future doubling of the levels of carbon dioxide in the atmosphere; scientific research which even our politicians are now realizing is of importance for our common future.

What seemed laughable and, perhaps, foolish to even the sophisticated and enlightened friends of Langley is today a unique tool for research in the physical sciences. This was a change in perception that occurred in a relatively short period of time. Doubt, questioning and the desire to see what, if anything is hidden under the noise that is flooding your detector when you make a measurement are far more important than absolute certainty for a scientist.

The greater the level of precision with which you examine, even a well-known phenomenon often leads to new physics, new discoveries and new insights into the nature of our physical world; and not simply to a more precise value of an already well-known constant. In this volume, we have looked at the recent changes made to the most widely used system of units, the SI. The SI has changed from being a system of units based on several 19th Century artefacts to the Quantum-SI, based on precise values of several constants of Nature. This change was only made possible by the work, over many years on improving the levels of precision with which measurements can be made by the international scientific community; for example, figure 14.1 shows a view into a research mass comparator at the BIPM, one of the most sensitive weighing devices ever constructed (we see the mass of a silicon sphere being compared to the mass of stainless-steel mass standards). In figure 14.2 we see another such silicon sphere (used to redefine the amount of substance, the mole) being manipulated—its surface perfection is clearly seen.

Some would say that science is organized doubt and that the scientist should never cease to question. Who knows what new worlds are waiting to be uncovered by increasing the level of precision and decreasing the degree of uncertainty with which we can measure a phenomenon, or by developing and improving the technology to make measurable what was once immeasurable? Today, we may think that we have made the ultimate measurement of a particular fundamental constant of Nature; the speed of light, the Planck constant, the Boltzmann constant or the mass and charge of the electron, but in a generation's time, new technology may well show that our previous measurements of these things were lacking and that we have to start again, but with a new paradigm about their interrelationship with the underlying laws of physics according to which the Universe unfolds.

14.2 The constants of Nature

We say that the Universe unfolds through the agency of interconnected universal laws and that these laws may be defined by associated constants of Nature such as

Figure 14.1. Looking inside a vacuum mass comparator (a sophisticated precision balance) where 1 kg mass prototypes from national metrology laboratories are weighed and calibrated. The electronic vacuum mass comparator is capable of determining differences in mass to an accuracy of 0.1 μg for weights of 1 kg under high-vacuum conditions (that is, 1 part in 10^{10}). In the pictures, we see a kilogram silicon sphere being manipulated (weighed relative to a traditional stainless-steel kilogram mass standard), and experiments investigating the mass of two silicon spheres, as part of the programme to create the Quantum-SI. (These images are reproduced with the permission of the BIPM, which retains full international copyright.)

the speed of light (c), the mass (m_e) and charge of the electron (e) or the Planck constant (h), see figure 8.1, but what exactly is a constant of Nature?

Consider the Planck constant. Max Planck found that the energy content of radiation (E) was proportional to the frequency of that radiation (ν). Proportionality does not, however, mean equality. A mathematician would write $E \propto \nu$; that is, the energy is proportional to the frequency. To go one step further and say that there is an equality, which is what is required if one wishes to use this relationship in theories and mathematical models, is to move from the concept of proportionality to mathematical equality. This change is accomplished by defining a constant of proportionality. Thus, we go on to write $E = (\text{constant})\,\nu$; or the energy of the radiation is equal to the frequency of the radiation multiplied by a constant, thereby giving us Planck's law or the Planck equation, $E = h\nu$. Max Planck investigated this particular constant and found its value and physical dimensions; and it is now

Figure 14.2. A highly polished, near-perfect 1 kg silicon sphere being manipulated during measurements at the Bureau internationale des poids et mesures (BIPM). The quality of the surface of this single-crystal of silicon may be seen in the detail of the reflection of the BIPM scientist, and behind him of the camera and photographer who took this photograph. These 1 kg, single-crystal silicon spheres (93.6 mm diameter) were manufactured for the Avogadro project and were used in defining the mol in the Quantum-SI (see section XX). These spheres are among the smoothest, round, man-made objects. If the surface of the sphere was scaled to the size of Earth, its highest point would rise to a maximum elevation of 2.4 m above the majority surface. (This image is reproduced with the permission of the BIPM, which retains full international copyright.)

known as the Planck constant, h, which relates the energy content of radiation to the frequency of that radiation, and which is now the basis of the definition of mass in the Quantum-SI (see chapter 13).

Such constants of physics are now at the heart of measurement science (see figure 8.1). This was foreseen by James Clerk Maxwell, who in his presidential address to the BAAS in 1870 said, 'If, then we wish to obtain standards of length, time, and mass which shall be absolutely permanent, we must seek them not in the dimensions, or the motion, or the mass of our planet, but in the wavelength, the period of vibration, and the absolute mass of these imperishable and unalterable and perfectly similar molecules'. The universal measure of Bishop John Wilkins, which became the Metric System's metre in 1795, was derived from a detailed survey of the distance from Dunkirk to Barcelona was Earth-specific. Maxwell wanted a universal measure, which was truly universal; a length standard which would be as valid here on Earth as it would be in another solar system. At that time, science and technology did not allow Maxwell's vision to be implemented; but today, the Quantum-SI is a fact, because many of the constants of Nature are known to levels of precision that would have astonished Maxwell and his generation.

14.3 Final thoughts on the evolution of units of measurement

By this point, I am sure the reader will have appreciated that creating a system of weights and measures, or even re-creating such a system to be acceptable to all is no easy thing. Even ignoring political conflicts, it is rarely possible to achieve consensus between relatively small groups of scientists as to which units they should be using.

And as for devising a scientifically coherent system of units that may be adopted and used by the wider society…there is, as you will have observed from these pages, no simple answer.

The creation of the Quantum-SI, will undoubtedly lead to new advances in many areas of basic science; particularly, in physics. But it will also have consequences on the teaching of the physical sciences; quantities that, before May 2019 were defined as constants of Nature are no longer constants of Nature in the new-SI (see section 13.3). A great deal of effort will be needed to communicate to the wider, non-scientific community the importance and details of the recent changes to the SI, but we are finally in a position to achieve the goal sought by two of the creators of the Metric System, Talleyrand and Condorcet; namely, to establish a system of units based on Nature that would be «naturel et invariable» and «ne renfermerait rien d'arbitraire ni de particulier à la situation d'aucun peuple sur le globe» (natural and unchanging … containing nothing arbitrary or referenced to a particular nation). With the transformation of the old-SI to the Quantum-SI, we have an appropriate vehicle for the great advances of science that are to come.

Ingram Content Group UK Ltd.
Milton Keynes UK
UKHW052351030523
421164UK00002B/3

9 780750 331418